# Cambridge Elements

Elements in The Philosophy of Mathematics
edited by
Penelope Rush
*University of Tasmania*
Stewart Shapiro
*The Ohio State University*

# MATHEMATICAL STRUCTURALISM

Geoffrey Hellman
*University of Minnesota*
Stewart Shapiro
*The Ohio State University*

CAMBRIDGE
UNIVERSITY PRESS

# CAMBRIDGE
## UNIVERSITY PRESS

University Printing House, Cambridge CB2 8BS, United Kingdom

One Liberty Plaza, 20th Floor, New York, NY 10006, USA

477 Williamstown Road, Port Melbourne, VIC 3207, Australia

314–321, 3rd Floor, Plot 3, Splendor Forum, Jasola District Centre,
New Delhi – 110025, India

79 Anson Road, #06–04/06, Singapore 079906

Cambridge University Press is part of the University of Cambridge.

It furthers the University's mission by disseminating knowledge in the pursuit of
education, learning, and research at the highest international levels of excellence.

www.cambridge.org
Information on this title: www.cambridge.org/9781108456432
DOI: 10.1017/9781108582933

First published 2019

*A catalogue record for this publication is available from the British Library.*

ISBN 978-1-108-45643-2 Paperback
ISSN 2399-2883 (online)
ISSN 2514-3808 (print)

# Mathematical Structuralism

DOI: 10.1017/9781108582933
First published online: 7 January 2019

Geoffrey Hellman
*University of Minnesota*

Stewart Shapiro
*The Ohio State University*

**Abstract:** The present work is a systematic study of five frameworks or perspectives articulating mathematical structuralism, whose core idea is that mathematics is concerned primarily with interrelations in abstraction from the nature of objects. The first two, set-theoretic and category-theoretic, arose within mathematics itself. After exposing a number of problems they encounter, Geoffrey Hellman and Stewart Shapiro consider three further perspectives formulated by logicians and philosophers of mathematics: sui generis, treating structures as abstract universals; modal, eliminating structures as objects in favor of freely entertained logical possibilities; and finally, modal-set-theoretic, a sort of synthesis of the set-theoretic and modal perspectives.

**Keywords:** mathematical structuralism, extendability of structures, logic of plurals, quasi-categoricity, assertory axioms

ISBNs: 9781108456432 (PB), 9781108582933 (OC)
ISSNs: 2399-2883 (online), 2514-3808 (print)

# Contents

# 1 Introduction

This section provides a sketch of structuralism in the philosophy of mathematics, focusing on features shared by all (or most) of the structuralist views in a wide philosophical context. We then provide a list of questions and criteria on which the various structuralist philosophies will be evaluated in subsequent sections.

## Overview

The theme of structuralism is that what matters to a mathematical theory is not the internal nature of its objects – numbers, functions, functionals, points, regions, sets, etc. – but how those objects relate to each other. The orientation grew from relatively recent developments within mathematics, notably toward the end of the nineteenth century and continuing through the present, particularly (but not exclusively) in the program of categorical foundations. Some of the relevant history is recounted in Section 3.

Mathematical structuralism is similar, in some ways, to functionalist views in, for example, philosophy of mind. A functional definition of a mental concept, such as *belief* or *desire*, is, in effect, a structural one, since it, too, focuses almost exclusively on relations that certain items have to each other. The difference is that mathematical structures are more abstract, and free-standing, in the sense that there are no restrictions on the kind of things that can exemplify them (see Shapiro 1997, Chapter 3, §6).

There are a number of mutually incompatible ways to articulate the structuralist theme, invoking various ontological and epistemic theses. Some philosophers postulate a robust ontology of structures, and their places, and then claim that the subject matter of a given branch of mathematics is a particular structure, or a class of structures. An advocate of a view like this would articulate what a structure is, and then say something about the metaphysical nature of structures, and how they and their properties can become known. There are also versions of structuralism amenable to those who deny the existence of distinctively mathematical objects altogether. And there are versions of structuralism in between, postulating an ontology for mathematics, but not a specific realm of structures.

Define a *system* to be a collection of objects together with certain relations on those objects. An extended family is a system of people under certain blood and marital relations – father, aunt, great niece, son-in-law, etc. A work of music is a collection of notes under certain temporal and other musical relations. To get closer to mathematics, define a *natural number system* to be a countably infinite collection of objects with a designated initial object, and a one-to-one successor relation that satisfies the axioms of second-order arithmetic (including the second-order induction axiom). Examples of natural number systems are the Arabic numerals in their natural order; a countably infinite sequence of distinct moments of time, say one second apart, in temporal order; the strings on a finite (or countable) alphabet arranged in lexical order; and, perhaps, the natural numbers themselves. Define a *Euclidean system* to be three collections of objects, one to be called "points," a second to be called "lines," and a third to be called "planes," along with certain relations between them, such that the Euclidean axioms are true of those objects and relations, so construed.

A *structure* is the abstract form of a system, which ignores or abstracts away from any features of the objects that do not bear on the relations. So the natural number structure is the form common to all of the natural number systems. The Euclidean structure is the form common to all Euclidean systems, etc.

A structure is thus a "one over many," a sort of universal. The main difference between a structure and a more traditional universal, such as a property, is that a given property applies to, or holds of, individual objects, while a given structure applies to, or holds of, entire *systems*.

Any of the usual array of philosophical views on universals can be adapted to structures, thus giving rise to some of the varieties of structuralism. One can be a Platonic *ante rem* realist about structures, holding that each structure exists and has its properties independent of any systems that have that structure – or at least independent of those systems that are not themselves structures. We call this view *sui generis structuralism* (SGS). On this view, structures exist objectively, and are ontologically prior to any systems that have them (or at least they are ontologically independent of such systems). Or one can be an Aristotelian *in re* realist, holding that structures exist, but insisting that they are ontologically posterior to the systems that instantiate them. One variety of this *in re* view is what we call *set-theoretic structuralism* (STS). On that view, structures are isomorphism types (or representatives thereof) within the set-theoretic hierarchy. The distinction between these two kinds of realism raises metaphysical issues of grounding and ontological priority.

Another option is to deny that structures exist at all. Talk of a given structure is just convenient shorthand for talk of all systems that are isomorphic to each other, in the relevant ways. Views like this are sometimes called *eliminative* structuralism, since they eschew the existence of structures altogether.

Advocates of the different ontological positions concerning structures take different approaches to other central philosophical concerns, such as epistemology, semantics, and methodology. Each such view has it relatively easy with some issues and finds deep, perhaps intractable problems with others. The *ante rem* SGS view, for example, has a straightforward account of reference and of the semantics of the languages of mathematics: the variables of a branch of mathematics, such as arithmetic, real analysis, or complex analysis, range over the places in an *ante rem* structure. Singular terms denote individual places, so the language is understood at face value.

In other words, an advocate of SGS has it that the straightforward grammatical structure of a mathematical language reflects the underlying logical form of the propositions. For example, in the simple arithmetic equation, $3 \times 8 = 24$, the numerals '3', '8', and '24' at least seem to be singular terms – proper names. In the SGS view, they *are* singular terms. The role of a singular term is to denote an individual object and, in the SGS view, each of these numerals denotes a place in the natural number structure. And, of course, the equation expresses a truth about that structure. In this respect, then, SGS is a variation on traditional Platonism. For this perspective to make sense, however, one has to think of a place in a structure as a *bona fide* object, the sort of thing that can be denoted by a singular term, and the sort of thing that can be in the range of first-order variables.

An advocate of the SGS approach agrees with the eliminativist (and the *in re* realist) that mathematical statements in, say, arithmetic, *imply* generalizations concerning systems that exemplify the structure. We say, for example, that in any natural number system, the object in the three-place multiplied (using the relevant relation in the system) by the object in the eight-place is the object in the twenty-four-place. Of course, the generalizations themselves do not entail that there are any natural number systems – nor any *ante rem* structures for that matter.

The eliminativist holds that mathematical statements *just are* (or are best interpreted as) generalizations like these, and she accuses the SG structuralist of making too much of their surface grammar, trying to draw deep metaphysical conclusions from that. For example, the simple theorem of arithmetic, "for every natural number $n$ there is a prime $p > n$" is rendered:

*In any natural number system S, for every object x in S, there is another object y in S such that y comes after x in S and y has no divisors in S other than itself and the unit object of S.*

In general, any sentence $\Phi$ in the language of arithmetic gets regimented as something like the following:

In any natural number system $S$, $\Phi[S]$, $\qquad\qquad\qquad\qquad (\Phi')$

where $\Phi[S]$ is obtained from $\Phi$ by restricting the quantifiers to the objects in $S$, and interpreting the non-logical terminology in terms of the relations of $S$.

In a similar manner, the eliminative structuralist paraphrases or regiments – and deflates – what seem to be substantial metaphysical statements, the very statements made by her SGS opponent. For example, "the number 2 exists" becomes "in every natural number system $S$, there is an object in the 2-place of $S$"; or "real numbers exist" becomes "every real number system has objects in its places." These statements are trivially true – analytic, if you will – not the sort of statements that generate heated metaphysical arguments.

However, the sailing is not completely smooth for the eliminativist. Suppose, for example, that the entire physical universe consists of no more than $10^{100,000}$ objects. Then there are no natural number systems (since each such system must have infinitely many objects). So for *any* sentence $\Phi$ in the language of arithmetic, the regimented sentence $\Phi'$ is vacuously true. So the eliminativist would be committed to the truth of (the regimented version of) $1 + 1 = 0$.

In other words, a straightforward, successful eliminative account of arithmetic requires a countably infinite background ontology. And it gets worse for other branches of mathematics. An eliminative account of real analysis demands an ontology whose size is that of the continuum; for functional analysis, we would need the power set of that many objects. And on it goes. The sizes of some of the structures studied in mathematics are staggering.

Even if the physical universe does exceed $10^{100,000}$ objects, and, indeed, even if it is infinite, there is surely *some* limit to how many physical objects there are. So, for the eliminative structuralist, branches of mathematics that, read at face value, require more objects than the number of physical objects end up being vacuously trivial. This would be bad news for such theorists, as the goal is to make sense of mathematics as practiced. In any case, no philosophy of mathematics should be hostage to empirical and contingent facts, including the number of objects in the physical universe.

There are two eliminativist reactions to this threat of vacuity. First, the philosopher might argue, or assume, that there are enough *abstract* objects for every mathematical structure to be exemplified. In other words, we postulate that, for each field of mathematics, there are enough abstract objects to keep the regimented statements from becoming vacuous.

Some mathematicians, and some philosophers, think of the set-theoretic hierarchy as the ontology for all of mathematics. Mathematical objects – all mathematical objects – are sets in the iterative hierarchy. Less controversially, it is often thought that the iterative hierarchy is rich enough to recapitulate every mathematical theory.

An eliminative structuralist might maintain that the theory of the background ontology for mathematics – set theory or some other – is not, after all, the theory of a particular structure. The foundation is a mathematical theory with an intended ontology in the usual, non-structuralist sense. In the case of set theory, the intended ontology is the sets. Set theory is not (merely) about all set-theoretic systems – all systems that satisfy the axioms. So the foundational theory is an exception to the theme of structuralism. But, the argument continues, every *other* branch of mathematics is to be understood in eliminative structuralist terms. This is the route of what we call STS.

Of course, this ontological version of eliminative structuralism is anathema to a nominalist, who rejects the existence of *abstracta* altogether. For the nominalist, sets and *ante rem* structures are pretty much on a par – neither is wanted. The other prominent eliminative reaction to the threat of vacuity is to invoke modality. In effect, one avoids (or attempts to avoid) a commitment to a vast ontology by inserting modal operators into the regimented generalizations. To reiterate the above example, the modal eliminativist renders "for every natural number $n$ there is a prime $p > n$" as something like:

*In any possible natural number system S, for every object x in S, there is another object y in S such that y comes after x in S and y has no divisors in S other than itself and the unit object of S.*

In general, let $\Phi$ be any sentence in the language of arithmetic; $\Phi$ gets regimented as:

In any possible natural number system $S, \Phi[S]$,

or, perhaps,

Necessarily, in any natural number system $S$, $\Phi[S]$,

where, again, $\Phi[S]$ is obtained from $\Phi$ by restricting the quantifiers to the objects in $S$, and interpreting the non-logical terminology in terms of the relations of $S$.

The modal structuralist also asserts that the various systems of mathematics are possible. It is possible for there to be a natural number system, a real number system, a Euclidean system, etc.

The difference with the ontological, eliminative program, of course, is that here the variables ranging over systems are inside the scope of a modal operator. So the modal eliminativist does not require an extensive, rich background ontology. Rather, she needs a large ontology to be possible.

The central problem with this brand of eliminativist structuralism concerns the nature of the invoked modality. Of course, it will not do much good to render the modality in terms of possible worlds. If one does that and takes possible worlds, and *possibilia*, to exist, then modal eliminative structuralism would collapse into the above ontological version of eliminative structuralism. Not much would be gained by adding the modal operators. The modalist typically takes the modality to be *primitive* – not defined in terms of anything more fundamental. But, of course, this move does not relieve the modalist of having to say something about the nature of the indicated modality, and something about how we know propositions about what is possible. We develop two versions of modal structuralism in subsequent sections, with references to the literature.

To briefly sum up and conclude, the parties to the debate over how best to articulate the structuralist insights agree that each of the major versions has its strengths and its peculiar difficulties. Negotiating such trade-offs is a stock feature of philosophy. The literature has produced an increased understanding of mathematics, of the relevant philosophical issues, and how the issues bear on each other.

## The State of the Economy

We plan to evaluate each version of structuralism by considering how well it fares on each of the following eight criteria.

(1)  What primitives are assumed and what is the background logic? Is it just first-order logic, or is second- or higher-order logic employed? If the latter, what is the status of relations and functions? What advantages and limitations are implied by these various choices?

(2)  The term "axioms" is ambiguous, as between "defining conditions on a type of structure of interest," on the one hand, and "basic assumptions" or "assertoric content," bearing a truth-value, on the other. It is characteristic of a structuralist view of mathematics to emphasize axioms in the former sense, as defining conditions on structures of interest; and this was the sense in which, for instance, Dedekind (1888) introduced the so-called "Peano postulates" on the natural number system in his classic essay, and it was axioms in this sense that Hilbert invoked in his well-known correspondence with Frege, who emphasized axioms in the assertoric sense (see the next section).

One should recognize that Frege had a point, viz. that a foundational framework requires some assertory axioms, capable of being true or false, governing especially the existence and nature of structures. When it comes to the different forms of structuralism, what are these "assertory" axioms?

(3) As an especially important case of (2), what assumptions are asserted as to the mathematical existence of structures? Is their indefinite extendability recognized or is there commitment to an absolutely maximal universe?

(4) Are structures recognized as a special type of object, or is there a thoroughgoing *elimination* of structures as objects? If not, what sort of objects are "structures," and, in particular, what is a *mathematical structure*?

(5) How is our epistemic access to structures understood, and what account of reference to them can be given?

(6) As an extension of (5), does the view allow for a face-value interpretation of mathematical statements? For example, do what appear to be singular terms in the languages of mathematics get rendered as singular terms? Do quantifiers get rendered as straightforward quantifiers? Or is there some regimentation or paraphrase involved?

(7) How are the paradoxes associated with set-theory and other foundational frameworks (such as category theory) to be resolved?

(8) Finally, how is Benacerraf's challenge based on competing identifications of numbers, etc., to be met?

## 2 Historical Background

Howard Stein (1988, p. 238) claims that during the nineteenth century, mathematics underwent "a transformation so profound that it is not too much to call it a second birth of the subject" – the first birth being in ancient Greece. The same period also saw important developments in philosophy, with mathematics as a central case study.

According to Alberto Coffa (1991, p. 7), for "better or worse, almost every philosophical development since 1800 has been a response to Kant." A main item on the agenda was to account for the *prima facie* necessity of mathematical propositions, the *a priori* nature of mathematical knowledge, and the applicability of mathematics to the physical world, all without invoking Kantian intuition. Can we understand mathematics independent of the forms of spatial and temporal intuition?

Coffa argues that the most successful approach to this problem was that of what he calls the " semantic tradition," running through the work of Bernard Bolzano, Gottlob Frege, the early Wittgenstein, and David Hilbert, culminating with the Vienna Circle, notably Moritz Schlick and Rudolf Carnap. The plan was to understand necessity and a priority in *formal* terms. In one way or another, this philosophical tradition was linked to the developments in mathematics. One legacy left by the developments in both mathematics and philosophy is mathematical logic, and model-theoretic semantics in particular. The emergence of model theory and the emergence of structuralism are, in a sense, the same.

In this section, we recount some themes in the development of Euclidean, projective, and non-Euclidean geometry, as well as some themes in arithmetic. Concerning geometry, there was a gradual transformation from the study of absolute or perceived space – matter and extension – to the study of structures. Our narrative includes sketches of early-twentieth-century theorists who either developed structuralist insights, or opposed these moves, or both. The list includes Dedekind, Frege, and Hilbert, among others.[1]

## Geometry, Space, Structure

The historical transition away from geometry as the study of physical or perceived space is complex. One early theme is the advent and success of analytic geometry, with projective geometry as a response. Another is the attempt to accommodate ideal and imaginary elements, such as points at infinity. A third thread is the assimilation of non-Euclidean geometry into mainstream mathematics (and into physics). These themes contributed to a growing interest in rigor and the eventual detailed understanding of rigorous deduction as independent of content – ultimately to a structuralist understanding of mathematics. Here, we can provide no more than a mere sketch of a scratch of this rich and wonderful history.

The traditional view of geometry is that its subject matter is matter and extension. The truths of geometry seem to be necessary, and yet geometry has something to do with the relations between physical bodies. Kant's account of geometry as synthetic *a priori*, relating to the forms of perceptual intuition, was a heroic attempt to accommodate the necessity, the *a priori* nature, *and* the empirical applicability of geometry.

The traditional view of arithmetic is that its subject matter is quantity. Arithmetic was the study of the discrete, while geometry was the study of the continuous. The fields were united under the rubric of mathematics, but one might wonder what they have in common other than this undescribed genus. The development of analytic geometry went some way toward loosening the

---

[1] Much of this section draws from Shapiro 1997, Chapter 5), used with kind permission from Oxford Universith Press, as well as Nagel (1939), Freudenthal (1962), Coffa (1986; 1991, Chapters 3 and 7), and Wilson (1992). Readers interested in these episodes of mathematical history are urged to consult those excellent works.

distinction between them. Mathematicians discovered that the study of quantity can shed light on matter and extension (and vice versa).

One result of the development of analytic geometry was that synthetic geometry, with its reliance on diagrams, fell into neglect. Joseph-Louis Lagrange even boasted that his celebrated treatise on mechanics did not contain a single diagram (but one might wonder whether his readers appreciated this feature). The dominance of analytic geometry left a void that affected important engineering projects. For example, problems with plane representations of three-dimensional figures were not tackled by mathematicians. The engineering gap was filled by the emergence of projective geometry (see Nagel 1939, §§7–8). Roughly, projective geometry concerns spatial relations that do not depend on fixed distances and magnitudes, nor on congruence. In particular, projective geometry dispenses with quantitative elements, like a metric.

Although all geometers continued to identify their subject matter as intuitable, visualizable figures in space, the introduction of so-called ideal elements, such as imaginary points, into projective geometry constituted an important move away from visualization. Parallel lines were thought to intersect, at a "point at infinity," although, of course, no one can visualize that, in any literal sense. Girard Desargues proposed that the conic sections – circle, ellipse, parabola, and hyperbola – form a single family of curves, since they are all projections of a common figure from a single "improper point" – located at infinity. Circles that do not intersect in the real plane were thought to have a pair of imaginary points of intersection. As Ernest Nagel (1939) put it, the "consequences for geometrical techniques were important, startling, and to some geometers rather disquieting" (p. 150). Clearly, mathematicians could not rely on the forms of perceptual intuition when dealing with the new imaginary elements. The elements are not in perceivable space; we do not *see* anything like them.

The introduction and use of imaginary elements in analytic and projective geometry were an outgrowth of the development of negative, transcendental, and imaginary numbers in arithmetic and analysis. With the clarity of hindsight, there are essentially three ways that "new" entities have been incorporated into mathematical theories (see Nagel 1979). One is to *postulate* the existence of mathematical entities that obey certain laws, most of which are valid for other, accepted entities. For example, one can think of complex numbers as like real numbers but closed under the taking of roots, and one can think of ideal points as like real points but not located in the same places. Of course, postulation begs the question against anyone who has doubts about the entities. Recall Bertrand Russell's ([1919] 1993, p. 71) quip about how postulation has the advantages of theft over honest toil.

In reply, one might point to the *usefulness* of the new entities, especially for obtaining results about established mathematical objects. But this benefit can be obtained with *any* system that obeys the stipulated laws. Thus, the second method is *implicit definition*. The mathematician gives a description of the system of entities, usually by specifying its laws, and then asserts that the description applies to any collection that obeys the stipulated laws. At this point, the skeptic might wonder whether there are any systems of entities that obey the stipulated laws.

The third method is construction, where the mathematician defines the new entities as combinations of already established objects. Presumably, this is the safest method since it settles the question of whether the entities exist (assuming the already-established objects do). William Rowan Hamilton's definition of complex numbers as pairs of real numbers fits this mold as does the logicist definition of natural numbers as collections of properties. A fruitful outlook would be to take implicit definition and construction in tandem. A construction of a system of objects establishes that there are systems of objects so defined, and so the implicit definition is not empty. Moreover, the construction also shows how the new entities can be related to the more established ones and may suggest new directions for research.

Nagel (1979) notes that all three methods were employed in the development of ideal points and points at infinity in geometry. Jean-Victor Poncelet came close to the method of postulation. In trying to explain the usefulness of complex numbers in obtaining results about the real numbers, he claimed that mathematical reasoning can be thought of as a mechanical operation with abstract signs. The results of such reasoning do not depend on any possible referents of the signs, so long as the rules are followed. Having thus "justified" new sorts of numbers in analysis, Poncelet went on to argue that geometry is equally entitled to employ abstract signs – with the same freedom from interpretation. He held that traditional, synthetic geometry is crippled by the insistence that everything be cast in terms of drawn or visualizable diagrams.

Poncelet's contemporaries were aware of the shortcomings of such bare postulation. Nagel cites authors like Joseph Diaz Gergonne and Hermann Grassmann, who more or less prefigured the method of implicit definition. Their work furthered the concern with rigor and the abandonment of the traditional view of geometry as concerned with extension. We move closer to a structuralist perspective on geometry. Grassmann's *Ausdehnungslehre* of 1844 developed geometry as "the general science of pure forms," considered in abstraction of any interpretation the language may have. He characterized the terms of geometry only by stipulated relations they have to each other:

No meaning is assigned to an element other than that. It is completely irrelevant what sort of specialization an element really is … it is also irrelevant in what respect one element differs from another, for it is specified simply as being different, without assigning a real content to the difference. (Grassmann, 1972, p. 47).

In this work, Grassmann dubbed the new study "the general science of pure forms" – to be considered independently of any intuitive content the theory might have. He distinguished "formal" from "real" sciences, along roughly the same lies as the contemporary distinction between pure and applied mathematics. Traditional geometry is a "real science" – an applied mathematics – and so *this* geometry

is not to be regarded as a branch of mathematics in the sense that arithmetic [is]. For [this] geometry refers to something given by nature (namely, space) and accordingly there must be a branch of mathematics which develops in an autonomous and abstract way laws which geometry predicates of space. [In this mathematics], all axioms expressing spatial intuitions would be entirely lacking … [T]he restriction that it be limited to the study of a three-dimensional manifold would … be dropped … Proofs in formal sciences do not go outside the domain of thought into some other domain … [T]he formal sciences must not take their point of departure from axioms, as do the real sciences, but will take definitions instead as their foundation. (ibid., pp. 10, 22)

Writing in 1877, almost thirty years later, Grassmann explained that his *Ausdehnungslehre* is

the abstract foundation of the doctrine of space … [I]t is free from all spatial intuition, and is a purely mathematical discipline whose application to space yields the structure of space. This latter science, since it refers to something given in nature (i.e., space) is no branch of mathematics, but is an application of mathematics to nature … For while geometry is limited to the three dimensions of space, [the] abstract science knows no such limitation. (ibid., p. 297)

As Nagel (1939, §36) put it, Grassmann was one of the first mathematicians who "explicitly recognized that mathematics is concerned with formal structures."

Concerning our third method of accommodating ideal elements, construction, Karl Georg Christian von Staudt (1856–60) showed that if we interpret the imaginary points as complex constructions of real points on real lines, then all of the theorems of the new projective geometry come out true. In modern terms, von Staudt discovered a model of the new theory. It is a particularly useful model since the supposed intended interpretation of original geometry – space – is a part of it.

A closely related development was the discovery of duality in projective geometry. We are accustomed, even today, to think of a line as a locus of points. However, one can just as well think of a point as a locus of lines. In projective geometry, with the ideal elements added, the symmetry between points and lines is deep. If the terminology for "points" and "lines" is systematically interchanged, all theorems still hold. Notice that with this duality, we contradict Euclid's definition of a point as that which has no parts. Interpreted via duality, the points do have parts – lines of the locus.

Impressed with this duality and extending it, Julius Plücker (1846, p. 322) wrote that geometric relations have validity "irrespective of every interpretation." When we prove a theorem concerning, say, straight lines, we have actually proved many theorems, one for each interpretation of the theory. The analogy with algebraic equations is apparent. Michel Chasles conceded that in light of the duality of projective geometry, the field should not be construed as the science of magnitude, but as the science of order (see Nagel 1939, §55, §59). The logical/structuralist insight is that, with sufficient rigor, the derivation of theorems should depend only on the stipulated relations between the elements and not on any features of this or that interpretation of them.

Whether or not primitive terms are to be understood in terms of spatial or temporal intuition, the emergence of rigor led to the idea that the inferences of a branch of mathematics should be independent of intuition. The longstanding idea that logic is a formal and topic-neutral canon of inference has concrete application in the present story. The topic-neutrality of logic dovetails with, and reinforces, the emancipation of geometry from matter and extension. Even if topic-neutrality had been expressed earlier in the history of mathematics and logic, it was carried out in some detail for geometry by Moritz Pasch (1926).[2] He made the aforementioned remark that when Euclid declared that a point has no parts, etc., he was not "explaining these concepts through properties of which any use can be made, and which in fact are not employed by him in the subsequent development" (pp. 15–16). Pasch thought it important that geometry be presented in a *formal* matter, without relying on intuition or observation when making inferences:

> If geometry is to be truly deductive, the process of inference must be independent in all its parts from the meaning of the geometrical concepts, just as it must be independent of the diagrams; only the relations specified in the propositions and definitions may legitimately be taken into account. During the deduction it is useful and legitimate, but in no way necessary, to think of the meanings of the terms; in fact, if it is necessary to do so, the inadequacy of

---

[2]   The first edition appeared in 1882 (Leipzig, Teubner).

the proof is made manifest. If, however, a theorem is rigorously derived from a set of propositions . . . the deduction has value which goes beyond its original purpose. For if, on replacing the geometric terms in the basic set of propositions by certain other terms true propositions are obtained, then corresponding replacements may be made in the theorem; in this way we obtain new theorems . . . without having to repeat the proof. (p. 91)

As suggested by this passage, this development was an important factor in the emergence of model theory as a major tool in mathematical logic. Pasch expresses the idea of geometry as a hypothetical-deductive endeavor, what he calls a "demonstrative science." Nagel (1939, §70) wrote that Pasch's work set the standard for contemporary (pure) geometry:

No work thereafter held the attention of students of the subject which did not begin with a careful enumeration of the undefined or primitive terms and unproved or primitive statements; and which did not satisfy the condition that all further terms be defined, and all further statements proved, solely by means of this primitive base.

This sounds much like the modern axiomatic method – and serves as a manifesto for structuralism.

Concerning the *source* of the axioms, however, Pasch held that geometry is a natural science – an *applied* mathematics. He was a straightforward, old-fashioned empiricist, holding that the axioms are verified by experience with bodies. We must wait for Henri Poincaré, Hilbert, and others to cut this final link with intuition or sensory experience. Pasch's empiricism died out, at least in mathematics; the need for formality and rigor did not.

Similar axiomatic programs came up in other branches of mathematics. In Italy, for example, Giuseppe Peano's influence led to a project to systematize all branches of pure mathematics. Mario Pieri, a member of this school, explicitly characterized geometry as a hypothetical-deductive enterprise. The subsequent development of *n*-dimensional geometry was but one fruit of this labor, and constituted another move away from intuition. Even if spatial intuition provides a little help in the heuristics of 4-dimensional geometry, intuition is an outright hindrance for 5-dimensional geometry and beyond.

Our final topic in this subsection is the emergence of non-Euclidean geometry. Here we see other steps toward a model-theoretic understanding of theories and thus another precursor to structuralism. We must be even more sketchy.

Michael Scanlan (1988) shows that the early pioneers of non-Euclidean geometry did not see themselves as providing models of uninterpreted axiom systems – in the present mold of model theory. These geometers were still thinking of their subject matter as physical or perceived space.

Eugenio Beltrami (1868a, 1868b, 1902), for example, took Euclidean geometry to be the true doctrine of space and, in that framework, he studied the plane geometry on various surfaces. "Lines" on surfaces are naturally interpreted as geodesic curves. In a flat space, like a Euclidean plane, the geodesic curves are the straight lines, and the geometry of such a plane is Euclidean. Beltrami showed how the geometries of some curved surfaces differ from Euclidean, thus moving toward non-Euclidean geometry. To be sure, Beltrami's procedure did involve *some* reinterpreting of the terms of geometry, and so it is a move in the direction of model theory and structuralism, but Scanlan shows that it is a small move. Beltrami could easily maintain that the terms all have their original meanings, transposed to the new contexts. "Line" means "geodesic curve," which is, perhaps, what it always meant anyway; "between" still means "between," etc. The meanings of the terms are not *as* fixed as they were in Euclid's *Elements*, but they are still spatial. It seems fair to think of the items in Beltrami's geometries as quasi-concrete in the sense of Parsons (1990) – they were all tied, in one way or another, to physical space. Eventually, of course, mathematicians began to consider more interpretations of the primitives of geometry, in less purely spatial contexts. The formal, or structural, outlook was the natural outcome of this process.

On most of the views available, one important scientific question was whether *physical* space is Euclidean. Since several theories of space were out in the open, the question of how one is to adjudicate them naturally arises. Which one describes the space we all inhabit? The surprising resolution of this question further aided the transition to a structuralist understanding of geometry.

Through the *Erlangen* program, Felix Klein (1921) made even more trouble for the question of which geometry is physically correct. Klein observed that the differences between Euclidean and non-Euclidean geometries lie in the different definitions of congruence, or the different measurements of distance, angle, and area. Each geometry can be imposed on the same domain of points. Moreover, Klein showed that with appropriate units, the numerical values of the different metrics need not differ appreciably from one another in sufficiently small neighborhoods. Consequently, if the issue of which geometry is correct still makes sense, it cannot be decided in practice. That is, even if we agree on the freely mobile, rigid measuring sticks, intuition and observation are not sufficiently precise to adjudicate the question of whether physical space is Euclidean. According to Klein, axiomatic systems introduce "exact statements into an inexact situation" (see Nagel 1939, §§81–82). His results show that our inexact intuitions are compatible with the assertion of different, technically incompatible, geometries: "Naive intuition is not exact, while refined intuition is not properly intuition at all, but rises through the logical development of axioms considered as perfectly exact." Thus,

> It is just at this point that I regard the non-Euclidean geometries as justified ... From this point of view it is a matter of course that of equally justifiable systems of axioms we prefer the simplest, and that we operate with Euclidean geometry just for that reason. (Klein 1921, p. 381)

A main theme of the *Erlangen* program was that different geometries can be characterized in terms of their "symmetries," or in other words, properties that are left fixed by certain isomorphisms. This focus on interpretation and isomorphism is of a piece with the rise of model theory and the emergence of structuralism.

Klein thus joined Pasch and a host of others in adopting an official empiricist philosophy of geometry even while his own work apparently undercut this philosophy. At this point, it seems, even the proponents of non-Euclidean geometry were not ready to drop intuition and perception entirely.

Hans Freudenthal (1962, p. 613) pointed out that the non-Euclidean developments were opposed on the ground that their advocates did not take intuition and perception seriously enough. In an amusing historical irony, some opponents dubbed the new science "meta-mathematics" or "meta-geometry." They meant these terms in a strong pejorative sense: non-Euclidean geometry is to respectable mathematics as metaphysics is to physics. It took some doing to exorcize intuition and perception entirely.

The advocates of non-Euclidean geometry won the war, at least in mathematics, but Euclidean geometry survived, albeit not as a theory of physical or perceived space (or space-time). The emerging status of Euclidean geometry, alongside the other geometries, makes for a nice philosophical case study in the emergence of structuralism.

With characteristic wit, Coffa (1986, pp. 8, 17) noted that the emergence of the different geometries posed a problem:

> During the second half of the nineteenth century, through a process still awaiting explanation, the community of geometers reached the conclusion that all geometries were here to stay ... [T]his had all the appearance of being the first time that a community of scientists had agreed to accept in a not-merely-provisory way all the members of a set of mutually inconsistent theories about a certain domain ... It was now up to philosophers ... to make epistemological sense of the mathematicians' attitude toward geometry ... The challenge was a difficult test for philosophers, a test which (sad to say) they all failed...

> For decades professional philosophers had remained largely unmoved by
> the new developments, watching them from afar or not at all . . . As the trend
> toward formalism became stronger and more definite, however, some phi-
> losophers concluded that the noble science of geometry was taking too harsh
> a beating from its practitioners. Perhaps it was time to take a stand on their
> behalf. In 1899, philosophy and geometry finally stood in eyeball-to-eyeball
> confrontation. The issue was to determine what, exactly, was going on in the
> new geometry.

Final blows to a role for intuition and perception in geometry were dealt by
Poincaré and Hilbert. Although Frege was himself an accomplished mathematician, he provided stiff resistance to the new geometry.

## Poincaré

Poincaré (1899, 1900) carried Klein's thread to its logical conclusion. With
Klein, Poincaré argued that it is impossible to figure out whether physical
space is Euclidean by an experiment, such as a series of measurements.
First, we have no access to space other than through configurations of
physical objects. Second, measurements can only be done upon physical
objects with physical objects. How can we tell whether the instruments
themselves conform to Euclidean geometry? For example, how can we tell
which edges are really straight? By further measurements? Clearly, there is
a vicious regress.

Poincaré then echoed Klein's conclusion that there is no fact about space
expressible in Euclidean terms that cannot be stated in any of the standard non-
Euclidean systems. The only difference is that things covered by one name (e.
g., "straight line") under one system would be covered by different names on
the second. Poincaré (1908, p. 235) wrote:

> We know rectilinear triangles the sum of whose angles is equal to two right
> angles; but equally we know curvilinear triangles the sum of whose angles is
> less than two right angles. The existence of the one sort is no more doubtful
> than that of the other. To give the name of straights to the sides of the first is to
> adopt Euclidean geometry; to give the name of straights to the sides of the
> latter is to adopt the non-Euclidean geometry. So that to ask what geometry is
> proper to adopt is to ask, to what line is it proper to give the name straight?

Poincaré noted that the various geometries are fully intertranslatable. Thus, the
choice of geometry is only a matter of a *convention*. All that remains is for the
theorist to stipulate which objects have straight edges and which do not. Space
itself has no metric; we impose one.

Poincaré held that it is meaningless to ask whether or not Euclidean geometry is *true*:

As well ask whether the metric system is true and the old measures false; whether Cartesian coordinates are true and polar coordinates false. One geometry cannot be more true than another; it can only be more convenient. (Poincaré 1908, p. 25)

Poincaré's view was that to adopt Euclidean geometry is to accept a series of complicated "disguised definitions" for specifying what sorts of configurations we will call "points," "lines," "triangles," etc. These "definitions" have at least a family resemblance to implicit definitions. Strictly speaking, the Euclidean "definitions" do not tell us what a point is. Instead, they specify how points are related to lines, triangles, etc. For Poincaré, words like "point" and "line" have no independent meaning given by intuition or perception. A point is anything that satisfies the conditions laid down by the axioms of the adopted geometry. Elsewhere (1900, p. 78), he wrote:

I do not know whether outside mathematics one can conceive a term independently of relations to other terms; but I know it to be impossible for the objects of mathematics. If one wants to isolate a term and abstract from its relations to other terms, what remains is nothing.

This looks like a structuralist insight, but we are not there yet. Poincaré still held that geometry is about (physical) space and that we have intuitions about this space. These intuitions, however, do not determine whether we should use Euclidean or non-Euclidean geometry. In modern terms, the intuitions under-determine the theory. The geometer adopts a convention concerning terms like "point" and "line." This convention determines which metric structure we are going to impose on space, and so the convention determines the geometry.[3]

As part of his conventionalism, Poincaré held that the theorems of geometry are not propositions with determinate truth values. As Coffa noted, this took some philosophers by surprise.

## Hilbert and the Emergence of Logic

The program executed in Hilbert's *Grundlagen der Geometrie* (1899) marked both an end to an essential role for intuition in geometry and the beginning of a

---

[3] Like Klein, Poincaré (1908, chapter 5) believed that Euclidean geometry would remain the most convenient theory, no matter what direction science took. With hindsight, the following passage is most ironic:

In astronomy "straight line" means simply "path of a ray of light." If therefore negative parallaxes were found ... two courses would remain open to us; we might either renounce Euclidean geometry, or else modify the laws of optics and suppose that light does not travel rigorously in a straight line. It is needless to add that all the world would regard the latter solution as the most advantageous.

fruitful era of meta-mathematics. Structuralism – in some form or other – is little more than a corollary to these developments.

Hilbert was aware that at some level, spatial intuition or observation remains the source of the axioms of geometry. In Hilbert's writing, however, the role of intuition is carefully and rigorously limited to motivation and heuristic. Once the axioms have been formulated, intuition is banished. It is no part of mathematics, whether pure or applied. The epigraph of Hilbert (1899) is a quotation from Kant's *Critique of Pure Reason* (A702/B730), "All human knowledge begins with intuitions, thence passes to concepts and ends with ideas," but the plan executed in that work is far from Kantian. In Hilbert's hands, the slogan "passes to concepts and ends with ideas" comes to something like "is *replaced* by logical relations between ideas." In the short "Introduction," he wrote:

> Geometry, like arithmetic, requires for its logical development only a small number of simple, fundamental principles. These fundamental principles are called the axioms of geometry. The choice of the axioms and the investigation of their relations to one another . . . is tantamount to the logical analysis of our intuition of space.

One result of the banishment of intuition is that the presented structure is free-standing: anything at all can play the role of the undefined primitives of points, lines, planes, etc., so long as the axioms are satisfied. Otto Blumenthal reports that in a discussion in a Berlin train station in 1891, Hilbert said that in a proper axiomatization of geometry, "one must always be able to say, instead of 'points, straight lines, and planes', 'tables, chairs, and beer mugs'."[4]

As noted above, Pasch and others had previously emphasized that the key to rigor is that our language and theorems be topic-neutral. With Hilbert, however, we see consequences concerning the essential nature of the very subject matter of mathematics. We also see an emerging meta-mathematical perspective of "logical analysis," with its own important meta-mathematical questions.

Hilbert (1899) does not contain the phrase "implicit definition," but the book clearly delivers implicit definitions of geometric structures. The early pages contain phrases like "the axioms of this group define the idea expressed by the word 'between' ..." and "the axioms of this group define the notion of congruence or motion." The idea is summed up as follows:

> We think of ... points, straight lines, and planes as having certain mutual relations, which we indicate by means of such words as "are situated," "between," "parallel," "congruent," " continuous," etc. The complete and

---

[4] " Lebensgeschichte" in Hilbert (1935, pp. 388–429); the story is related on p. 403. See Stein (1988, p. 253), Coffa (1991, p. 135), and Hallett (1990, pp. 201–2).

exact description of these relations follows as a consequence of the axioms of geometry.

To be sure, Hilbert also says that the axioms express "certain related funda-mental facts of our intuition," but in the subsequent development of the book, all that remains of the intuitive content is the use of words like "point," "line," etc., and the diagrams that accompany some of the theorems.

Paul Bernays (1967, p. 497) sums up the aims of Hilbert (1899):

> A main feature of Hilbert's axiomatization of geometry is that the axiomatic method is presented and practiced in the spirit of the abstract conception of mathematics that arose at the end of the nineteenth century and which has generally been adopted in modern mathematics. It consists in abstracting from the intuitive meaning of the terms ... and in under-standing the assertions (theorems) of the axiomatized theory in a hypothetical sense, that is, as holding true for any interpretation ... for which the axioms are satisfied. Thus, an axiom system is regarded not as a system of statements about a subject matter but as a system of condi-tions for what might be called a relational structure ... [On] this con-ception of axiomatics, ... logical reasoning on the basis of the axioms is used not merely as a means of assisting intuition in the study of spatial figures; rather logical dependencies are considered for their own sake, and it is insisted that in reasoning we should rely only on those proper-ties of a figure that either are explicitly assumed or follow logically from the assumptions and axioms.

This suggests an *eliminative* structuralism. A mathematical theory is about *any* system of objects that satisfies the axioms or defining conditions.

Interest in meta-mathematical questions surely grew from the developments in non-Euclidean geometry, as a response to the failure to prove the parallel postulate. In effect (and with hindsight), the axioms of non-Euclidean geometry were shown to be satisfiable. Hilbert (1899) raised and solved questions concerning the satisfiability of sets of geometric axioms. Using techniques from analytic geometry, he constructed a model of all of the axioms using real numbers, thus showing that the axioms are "compatible," or satisfiable. If spatial intuition were playing a role beyond heuristics, this proof would not be necessary. Intuition alone would assure us that all of the axioms are true (of real space), and thus that they are all compatible with each another. Geometers in Kant's day would wonder what the point of this exercise is. As we shall see, Frege also balked at it.

Hilbert then gave a series of models in which one axiom is false, but all the other axioms hold, thus showing that the indicated axiom is independent of the others. The various domains of points, lines, etc. of each model are sets of numbers, sets of pairs of numbers, or sets of sets of numbers. Predicates and relations, such as "between" and "congruent," are interpreted over the given domains in the now familiar manner.

The second of Hilbert's famous "Problems" (Hilbert 1900) extends the meta-mathematical approach to every corner of mathematics:

> When we are engaged in investigating the foundations of a science, we must set up a system of axioms which contains an exact and complete description of the relations subsisting between the elementary ideas of that science. The axioms set up are at the same time the definitions of those elementary ideas.

Once again, the "definition" is an implicit definition.

It is easy to see that isomorphic models are equivalent. If there is a one-to-one function from the domain of one model onto the domain of the other that preserves the relations of the model, then any sentence of the formal language that is true in one model is true in the other. So, if intuition or perception via ostension is banned, then the best a formal theory can do is to fix its interpretation up to isomorphism. Any model of the theory can be changed into another model just by substituting elements of the domain and renaming.

Hermann Weyl (1949, pp. 25–27) put it well:

> [A]n axiom system [is] a logical mold of possible sciences ... One might have thought of calling an axiom system complete if in order to fix the meanings of the basic concepts present in them it is sufficient to require that the axioms to be valid. But this ideal of uniqueness cannot be realized, for the result of an isomorphic mapping of a concrete interpretation is surely again a concrete interpretation ... A science can determine its domain of investigation up to an isomorphic mapping. In particular, it remains quite indifferent as to the "essence" of its objects ... The idea of isomorphism demarcates the self-evident boundary of cognition ... Pure mathematics ... develops the theory of logical "molds" without binding itself to one or the other among possible concrete interpretations ... The axioms become implicit definitions of the basic concepts occurring in them.

This is yet another structuralist manifesto, again suggesting an eliminative approach.

Of course, Hilbert's meta-mathematics did not come out of the blue. His contemporaries included the Italians Alessandro Padoa, Pieri, and Peano. Here, there is no need to elaborate the similarities and differences, nor to join arguments of priority. Hilbert's influence was due to the clarity and depth of his work. He led by example. Freudenthal (1962, pp. 619, 621) sums things up:

> The father of rigor in geometry is Pasch. The idea of the logical status of geometry occurred at the same time to some Italians. Implicit definition was

analyzed much earlier by Gergonne. The proof of independence by counter-example was practiced by the inventors of non-Euclidean geometry, and more consciously by Peano and Padoa ... [I]n spite of all these historical facts, we are accustomed to identify the turn of mathematics to axiomatics with Hilbert's *Grundlagen*: This thoroughly and profoundly elaborated piece of axiomatic workmanship was infinitely more persuasive than programmatic and philosophical speculations on space and axioms could ever be ... There is no clearer evidence for the persuasiveness of Hilbert's *Grundlagen*, for the convincing power of a philosophy that is not preached as a program, but that is only the silent background of a masterpiece of workmanship.

## Frege vs. Hilbert

Here we briefly recapitulate the spirited correspondence between Frege and Hilbert.[5] Frege opened the proceedings by lecturing Hilbert on the nature of definitions and axioms. According to Frege, axioms should express truths and definitions should give the meanings and fix the denotations of certain terms. With a Hilbert-style implicit definition, *neither* job is accomplished. In a letter dated December 27, 1899, Frege complained that Hilbert (1899) does not provide a definition of, say, "between" since the axiomatization "does not give a characteristic mark by which one could recognize whether the relation Between obtains":

> [T]he meanings of the words "point," "line," "between" are not given, but are assumed to be known in advance ... [I]t is also left unclear what you call a point. One first thinks of points in the sense of Euclidean geometry, a thought reinforced by the proposition that the axioms express fundamental facts of our intuition. But afterwards you think of a pair of numbers as a point ... Here the axioms are made to carry a burden that belongs to definitions ... [B]eside the old meaning of the word "axiom," ... there emerges another meaning but one which I cannot grasp. (Frege 1980, pp. 35–36)

Frege went on to remind Hilbert that a definition should specify the meaning of a single word whose meaning has not yet been given, and the definition should employ other words whose meanings are already known. In contrast to definitions, axioms and theorems

> must not contain a word or sign whose sense and meaning, or whose contribution to the expression of a thought, was not already completely laid down, so that there is no doubt about the sense of the proposition and the thought it expresses. The only question can be whether this thought is true and what its truth rests on. Thus axioms and theorems can never try to

---

[5]  See, for example, Resnik (1980), Coffa (1991, Ch. 7), Demopoulos (1994), and Hallett (1994). The Frege-Hilbert correspondence is published in Frege (1976) and translated in Frege (1980).

lay down the meaning of a sign or word that occurs in them, but it must
already be laid down. (Frege 1980, p. 36)

Frege's point is a simple one, and has considerable intuitive appeal. If the terms
in the proposed axioms do not have meaning beforehand, then the statements
cannot be true (or false), and thus they cannot be axioms. If, on the other hand,
the terms do have meaning beforehand, then the axioms cannot be definitions.
Frege added that from the truth of axioms, "it follows that they do not contradict one another" and so there is no further need to show that the axioms are
consistent.

   Hilbert replied just two days later, on December 29. He told Frege that the
purpose of Hilbert (1899) is to explore logical relations among the principles of
geometry, to see why the "parallel axiom is not a consequence of the other
axioms," and how the fact that the sum of the angles of a triangle is two right
angles is connected with the parallel axiom. Frege, the pioneer in mathematical
logic, could surely appreciate *this* project, or at least this aspect of the
Hilbertian project. The key lies in how Hilbert understood the logical relations.
Concerning Frege's assertion that the meanings of the words "point," "line,"
and "plane" are "not given, but are assumed to be known in advance," Hilbert
replied:

   This is apparently where the cardinal point of the misunderstanding lies. I do
   not want to assume anything as known in advance. I regard my explanation
   . . . as the definition of the concepts point, line, plane . . . If one is looking for
   other definitions of a "point," e.g. through paraphrase in terms of extension-
   less, etc., then I must indeed oppose such attempts in the most decisive way;
   one is looking for something one can never find because there is nothing
   there; and everything gets lost and becomes vague and tangled and degen-
   erates into a game of hide and seek. (Frege 1980, p. 39)

This is an allusion to definitions like Euclid's "a point is that which has no
parts," which were noted above. To try to do better than a characterization up to
isomorphism is to lapse into "hide and seek." Later in the same letter, when
responding to the complaint that his notion of "point" is not "unequivocally
fixed," Hilbert wrote:

   [I]t is surely obvious that every theory is only a scaffolding or schema of
   concepts together with their necessary relations to one another, and that the
   basic elements can be thought of in any way one likes. If in speaking of my
   points, I think of some system of things, e.g., the system love, law, chimney-
   sweep . . . and then assume all my axioms as relations between these things,
   then my propositions, e.g., Pythagoras' theorem, are also valid for these
   things . . . [A]ny theory can always be applied to infinitely many systems of
   basic elements. One only needs to apply a reversible one-one transformation

and lay it down that the axioms shall be correspondingly the same for the transformed things. This circumstance is in fact frequently made use of, e.g. in the principle of duality . . . [This] . . . can never be a defect in a theory, and it is in any case unavoidable. (Frege 1980, pp. 40–41)

Note the similarity with Hilbert's quip in the Berlin train station. Here it is elaborated in terms of isomorphism and logical consequence.

Since Hilbert did "not want to assume anything as known in advance," he rejected Frege's claim that there is no need to worry about the consistency of the axioms, because they are all true:

> As long as I have been thinking, writing and lecturing on these things, I have been saying the exact reverse: if the arbitrarily given axioms do not contradict each other with all their consequences, then they are true and the things defined by them exist. This is for me the criterion of truth and existence. (Frege 1980, pp. 39–40)

Hilbert then repeated the role of what is now called "implicit definition," noting that it is impossible to give a definition of "point" in a few lines since "only the whole structure of axioms yields a complete definition."

Frege's response (p. 43), dated January 6, 1900, acknowledged at least part of Hilbert's project. Frege saw that Hilbert wanted "to detach geometry from spatial intuition and to turn it into a purely logical science like arithmetic."[6] Frege also understood Hilbert's model-theoretic notion of consequence: to show that a sentence $D$ is not a consequence of $A$, $B$, and $C$ is to show that "the satisfaction of $A$, $B$, and $C$ does not contradict the non-satisfaction of $D$." However, these great minds were not to find much more in the way of common ground. Frege said that the only way to establish consistency is to give a model: "to point to an object that has all those properties, to give a case where all those requirements are satisfied."

In the same letter, Frege (p. 46) mocked implicit definitions, suggesting that with them, we can solve problems of theology:

> What would you say about the following?
> Explanation. We imagine objects we call Gods.

---

[6] Frege's logicism did not extend to geometry, which he regarded as synthetic *a priori*. To return to a theme broached above, Frege worried about the ideal elements introduced into geometry. In his inaugural dissertation (of 1873) he wrote:

> When one considers that the whole of geometry ultimately rests upon axioms, which receive their validity from the nature of our faculty of intuition, then the question of the sense of imaginary figures appears to be well-justified, since we often attribute to such figures properties which contradict our intuition. (Frege 1967, p. 1)

See Kitcher (1986, p. 300) and Wilson (1992).

Axiom 1. All Gods are omnipotent.

Axiom 2. All Gods are omnipresent.

Axiom 3. There is at least one God.

Frege did not elaborate and, as far as we know, Hilbert never responded to this. Frege's point is clear enough. Hilbert had said that if we establish the consistency of his axiomatization, we *thereby* establish the existence of points, lines, and planes. If we then establish the consistency of the theology axioms, do we thereby establish the existence of a God?

The difference, we presume, is that theology is not mathematics; it is not to be understood formally, in terms of the relations of the basic elements to each other. Not just anything can play the God-role, and the central properties of omnipotence and omnipresence are not formal.

Frege complained that Hilbert's "system of definitions is like a system of equations with several unknowns." This analogy is apt, and we think that Hilbert would accept it. Frege wrote: "Given your definitions, I do not know how to decide the question whether my pocket watch is a point." Again, Hilbert would surely agree. Presumably, Frege's pocket watch is indeed an element of a model of Euclidean geometry. To state the obvious, Frege did not think in terms of "truth in a model." For Frege, the quantifiers of mathematics range over everything, and a concept is a function that takes all objects as arguments. Thus, "my pocket watch is a point" must have a truth value, and our theory must determine this truth value.

Neither of them budged. On September 16 of that year, a frustrated Frege wrote that he could not reconcile the claim that axioms are definitions with Hilbert's view that axioms contain a precise and complete statement of the relations among the elementary concepts of a field of study (from Hilbert 1900, quoted above). For Frege, "there can be talk about relations between concepts … only after these concepts have been given sharp limits, but not while they are being defined."

On September 22, an exasperated Hilbert replied:

> [A] concept can be fixed logically only by its relations to other concepts. These relations, formulated in certain statements I call axioms, thus arriving at the view that axioms … are the definitions of the concepts. I did not think up this view because I had nothing better to do, but I found myself forced into it by the requirements of strictness in logical inference and in the logical construction of a theory. I have become convinced that the more subtle parts of mathematics … can be treated with certainty only in this way; otherwise one is only going around in a circle. (p. 51)

This, too, will do nicely as a structuralist manifesto.

Hilbert took the rejection of Frege's perspective on concepts to be a major innovation. In a letter to Frege dated November 7, 1903, he wrote that

> [T]he most important gap in the traditional structure of logic is the assumption ... that a concept is already there if one can state of any object whether or not it falls under it ... [Instead, what] is decisive is that the axioms that define the concept are free from contradiction. (pp. 51–52)

The opening paragraphs of Hilbert (1905) contain a similar point, explicitly directed against Frege.

For his part, Frege used his own logical system to recapitulate much of Hilbert's orientation. In the January 6 letter, Frege suggests that Hilbert's axiomatization does not provide an answer

> to the question " What properties must an object have in order to be a point (a line, a plane, etc.)?", but they contain, e.g., second-level relations, e.g., between the concept point and the concept line. (p. 46)

Frege was correct here, and Hilbert would surely agree. Hilbert did not give necessary and sufficient conditions for an arbitrary object to be a point. Instead, he showed how points are related to each other and to lines, planes, etc. He gave necessary and sufficient conditions for a *system* of objects to exemplify the relevant structure.

In subsequent lectures on geometry (1903b, 1906, translated in Frege 1971), Frege formulated these second-level relations with characteristic rigor. He replaced the geometric terms – the items we would call non-logical – with bound higher-order variables. In Frege's hands, Hilbert's implicit definition is transformed into an explicit definition of a *second-level concept*. Frege also used his logical system to recapitulate something like Hilbert's model-theoretic notion of independence. An assertion in the form "the satisfaction of *A, B,* and *C* does not contradict the non-satisfaction of *D*" is rendered as a formula in Frege's system.

Thus, it is not quite true that model theory was foreign to Frege's logical system. Since his language was designed to be universal, *anything* could be expressed in it – even statements about models of axiomatizations with noted non-logical terminology.

In sum, Frege and Hilbert did manage to understand each other, at least up to a point. Nevertheless, they were at cross-purposes in that neither of them saw much value in the other's point of view. The central feature of Hilbert's orientation is that the main concepts being defined are given different extensions in each model. This is the contemporary notion of a non-logical term (see Demopoulos 1994). Frege's own landmark work in logic, the *Begriffsschrift*

(1879), contains nothing like non-logical terminology. Every lexical item has a fixed sense and a fixed denotation or extension. Everything is already fully interpreted, and there is nothing to vary from interpretation to interpretation. This feature, shared by the other major logicist Russell, has been called "logic as language," whereas Hilbert's approach is "logic as calculus" (see, for example, van Heijenoort 1967b and Goldfarb 1979).

## Dedekind

Another logicist, Richard Dedekind, played a central role in the development of the structuralist perspective. His *Stetigkeit und irrationale Zahlen* (1872) contains the celebrated account of continuity and the real numbers. The opening paragraph sounds the now familiar goal of eliminating appeals to intuition:

> In discussing the notion of the approach of a variable magnitude to a fixed limiting value ... I had recourse to geometric evidences. Even now such resort to geometric intuition in a first presentation of the differential calculus I regard as extremely useful ... But that this form of introduction into the differential calculus can make no claim to being scientific no one will deny ... The statement is so frequently made that the differential calculus deals with continuous magnitude, and yet an explanation of this continuity is nowhere given. (opening paragraph, before §1)

Dedekind motivated the problem by noting that the *rational* numbers can be mapped, one-to-one, into any straight line, given only a point on the line as origin and an interval as unit. Thus, the arithmetical property of "betweenness" shares many (structural) features with its geometrical counterpart. Of course, the ancient lesson of incommensurables is that there are points on the line that do not correspond to any rational number. In other words, there are holes or gaps in the image of the rational numbers under any embedding. Consequently, the rational numbers are not continuous. So what is it to *be* continuous?

According to Dedekind, even if an appeal to spatial or temporal intuition were allowed at this point, it would not help. Do we have intuitions of continuity? Rudolf Lipschitz wrote to Dedekind that the property of continuity is self-evident and so does not need to be stated, since no one can conceive of a line as discontinuous. Dedekind replied that he, for one (and Cantor, for two) *can* conceive of discontinuous lines (Dedekind 1932, Volume 3, p. 478; see Stein 1988, p. 244). Dedekind (1872, §3) wrote that "if space has at all a real existence it is *not* necessary for it to be continuous," and in the preface to the first edition of his later monograph on the natural numbers (1888) he gave a nowhere continuous interpretation of Euclidean space. His point is that we do not *see* or *intuit* the line as continuous; we *attribute* continuity to it. But we have as yet no rigorous formulation of what continuity is. He sought an "axiom by which we attribute to the line its continuity."

Dedekind defined a cut to be a division of the rational numbers into two non-empty sets $(A_1, A_2)$ such that every member of $A_1$ is less than every member of $A_2$. If there is a rational number $n$ such that $n$ is either the largest member of $A_1$ or the smallest member of $A_2$, then the cut $(A_1A_2)$ is "produced" by $n$. Of course, some cuts are not produced by any rational number – thus the discontinuity of the rational numbers. Dedekind's plan was to fill the gaps:

> Whenever, then, we have to do with a cut $(A_1A_2)$ produced by no rational number, we create a new, an *irrational* number $a$, which we regard as completely defined by this cut $(A_1A_2)$; we shall say that the number $a$ corresponds to this cut, or that it produces this cut. (§4)

Dedekind's language (1872) suggests that did not *identify* the real numbers with the cuts. Instead, the "created" real numbers "correspond" to the cuts. It is not clear what Dedekind meant by the phrase "create a new number"; we shall return to this shortly.

The opening paragraphs of Dedekind's later monograph on the natural numbers (1888) continue the rejection of intuition:

> In speaking of arithmetic (algebra, analysis) as a part of logic I mean to imply that I consider the number-concept entirely independent of the notions or intuitions of space and time. (Preface)

This echoes a passage in the earlier work (1872, §1):

> I regard the whole of arithmetic as a necessary, or at least natural, consequence of the simplest arithmetic act, that of counting, and counting itself as nothing else than the successive creation of the infinite series of positive integers in which each individual is defined by the one immediately preceding; the simplest act is the passing from an already-formed individual to the consecutive new one to be formed.

The "counting" here is not the technique of determining the cardinality of a collection or concept extension by putting them in one-to-one correspondence with an initial segment of the numerals. Dedekind's counting is more like what Paul Benacerraf (1965) calls "intransitive counting," reciting names for the natural numbers in order. This is what corresponds to the "creation" of the natural numbers one after another.

Dedekind (1888, §5) defined a set $S$ and function $s$ to be a "simply infinite system" if $s$ is one-to-one, there is an element $e$ of $S$ such that $e$ is not in the range of $s$ (thus making $S$ Dedekind-infinite), and the only subset of $S$ that both contains $e$ and is closed under $s$ is $S$ itself. In effect, a simply infinite system is a model of the natural numbers, under the successor relation.

At this point, Dedekind defined the "natural numbers" with language that is music to the ears of a structuralist:

> If in the consideration of a simply infinite system $N$ set in order by a transformation $\varphi$ we entirely neglect the special character of the elements, simply retaining their distinguishability and taking into account only the relations to one another in which they are placed by the order-setting transformation $\varphi$, then are these numbers called *natural numbers* or *ordinal numbers* or simply *numbers* ... With reference to this freeing the elements from every other content (abstraction) we are justified in calling numbers a free creation of the human mind. The relations or laws which are derived entirely from the conditions ... are always the same in all ordered simply infinite systems, whatever names may happen to be given to the individual elements. (§6)

Like Frege (1884), Dedekind went on to establish that "the natural numbers" satisfy the usual arithmetic properties, such as the induction principle. He defined addition and multiplication on "the natural numbers" and proved that definitions by primitive recursion do define functions. Finally, Dedekind gave straightforward applications of "the natural numbers" to the counting and ordering of finite classes.

The last sentence quoted just above suggests a reading of Dedekind along the lines of what is today called "eliminative structuralism," a philosophy that eschews the existence of structures (and their places) as *bona fide* mathematical entities. Talk of "the natural numbers" is just a disguised way of talking about all simply infinite systems, all collections of objects with a given function with given properties. Charles Parsons (1990) proposes, but eventually drops, such an interpretation (see also Kitcher 1986, e.g., p. 333 and Mayberry 2000). Dedekind's published works (1872, 1888) are more or less neutral on the distinction between eliminative and sui generis (*ante rem*) structuralism; either interpretation fits most of the text. Dedekind did use the letter "$N$" as a singular term and there are phrases like "we denote numbers by small italics" (§124), which suggests a sui generis reading, but Dedekind's locution could mean something like "in what follows, we use '$N$' to denote an arbitrary (but unspecified) simply infinite system and we use small italics to denote the members of $N$." It is similar to a contemporary algebraist saying "let $G$ be an arbitrary group and let $e$ be the identity element of $G$."

The theorem proved in Dedekind (1888, §132) is "All simply infinite systems are similar [i.e., isomorphic] to the number series $N$ and consequently ... also to one another." On an eliminative reading, this is awkward and redundant. It would be better to say directly (and prove directly) that all simply infinite systems are isomorphic to each other. The same goes for other passages, like that in §133, but again, this does not rule out an eliminative reading. To further understand Dedekind's views on these matters, we turn to his correspondence.

Dedekind's friend Heinrich Weber suggested to him that real numbers be *identified* with Dedekind cuts, and that natural numbers – finite cardinals – be identified with classes, along the then-familiar lines of Frege or Whitehead and Russell (1910). Dedekind replied that there are many properties of cuts that would sound very odd if applied to the corresponding real numbers (Dedekind 1932, Vol. 3, pp. 489–490). Similarly, there are many properties of the class of all triplets that should not be applied to the number 3. This is essentially the same point made in Benacerraf's (1965) celebrated work.

In the same letter, Dedekind (1932) again invoked the mind's creativity:

> If one wished to pursue your way – and I would strongly recommend that this be carried out in detail – I should still advise that by number ... there be understood not the class (the system of all mutually similar finite systems), but rather something *new* (corresponding to this class), which the mind *creates*. We are of divine species and without doubt possess creative power not merely in material things ... but quite specially in intellectual things. This is the same question of which you speak ... concerning my theory of irrationals, where you say that the irrational number is nothing else than the cut itself, whereas I prefer to create something new (different from the cut), which corresponds to the cut ... We have the right to claim such a creative power, and besides it is much more suitable, for the sake of the homogeneity of all numbers, to proceed in this manner. (pp. 489–490)

Thus, for Dedekind, numbers *are* objects, in line with sui generis structuralism.[7]

As noted above, it is not clear what Dedekind meant by the "creativity" metaphor. Matters of charity preclude attributing some sort of subjectivism to this great mathematician. Dedekind surely did not think that the existence of the real numbers began at the moment he had the idea of "creating" them from cuts.

---

[7] Nevertheless, Dedekind did not insist on thinking of numbers as objects. In a letter to Lipschitz (June 10, 1876, Dedekind (1932, §65)), he wrote that "if one does not want to introduce new numbers, I have nothing against this; the theorem ... proved by me will now read: the system of all cuts in the discontinuous domain of all rational numbers forms a continuous manifold." This alternate reading is consonant with eliminative structuralism. Michael Hallett (1990, p. 233) remarks that Dedekind did not insist on numbers being objects because the existence of numbers as objects is not part of the meaning of "real number." What does matter is "the holding of the right properties," i.e., the structure. It is convenient and perspicuous to take numbers to be new objects, but such a view is not required.

Mathematics is objective if anything is; Dedekind's numbers are the same as anybody's, or at least they are the same as any mathematician's.[8]

Some commentators, like Charles McCarty (1995), suggest a Kantian reading of Dedekind's "creativity," relating it to the categories rational beings invoke in order to think. Stein (1988) relates "free creation" to the mind's ability to open and explore new conceptual possibilities, a theme that occurs often in Dedekind's writings.

The most common interpretation is that Dedekind's "free creation" is an abstraction of sorts. William Tait, for example, wrote of "Dedekind abstraction." Starting with, say, the finite von Neumann ordinals,

> we may ... abstract from the nature of these ordinals to obtain the system $N$ of natural numbers. In other words, we introduce $N$ together with an isomorphism between the two systems. In the same way, we can introduce the continuum, for example, by Dedekind abstraction from the system of Dedekind cuts. In this way, the arbitrariness of this or that particular "construction" of the numbers or the continuum ... is eliminated. (Tait 1986, p. 369)

If two systems are isomorphic, then the structures obtained from them by Dedekind abstraction are identical. This highlights the importance of the categoricity proof in Dedekind (1888), and explains why he had no objection to other ways of defining the real and natural numbers, and even encouraged that alternate paths be pursued. Dedekind (1888) in the preface agreed that Cantor and Weierstrass have both given "perfectly rigorous," and so presumably acceptable, accounts of the real numbers, even though they were different from his own and from each other. What the accounts share is the abstract structure.

In a letter to Hans Keferstein (February 1890; translated in van Heijenoort 1967a, pp. 99–103), Dedekind reiterated that he regarded the natural numbers to be a *single* collection of objects. He wrote that the "number sequence $N$ is a *system* of individuals, or elements, called numbers. This leads to the general consideration of systems as such." He called the number sequence $N$ the "abstract type" of a simply infinite system. In response to Keferstein's complaint that Dedekind had not given an adequate definition of the number 1, he wrote:

---

[8] In addition, Dedekind was no constructivist. He (1888, §1) explicitly adopted excluded middle, adding a footnote: "I mention this expressly because Kronecker not long ago ... endeavored to impose certain limitations on the free formation of concepts in mathematics which I do not believe to be justified."

I define the *number* 1 as the basic element of the number sequence without any ambiguity ... and just as unambiguously, I arrive at the *number* 1 ... as a consequence of the general definition ... Nothing further may be added to this at all if the matter is not to be muddled. (p. 102)

Note the similarity of the last remark to Hilbert's quip that any attempt to get beyond structural properties is to go around in a circle.

Georg Cantor also spoke of a process of abstraction, which produces finite and transfinite cardinal numbers as individuals. His language is similar to Dedekind's:

By the "power" or "cardinal number" of *M* we mean the general concept, which arises with the help of our active faculty of thought from the set *M*, in that we abstract from the nature of the particular elements of *M* and from the order they are presented ... [E]very single element *m*, if we abstract from its nature, becomes a "unit"; the cardinal number [of *M*] is a definite aggregate composed of units, and this number has existence in our mind as an intellectual image or projection of the given aggregate *M*. (Cantor 1932, pp. 282–3)

Cantor gave a similar account of order types, starting with an ordered system:

[W]e understand the general concept which arises from *M* when we abstract only from the nature of the elements of *M*, retaining only the order of precedence among them ... Thus the order type ... is itself an ordered set whose elements are pure units. (Cantor 1932, p. 297)

Cantor thus shared a notion of "free creation" – abstraction – with Dedekind. Russell (1903, p. 249) took both Cantor and Dedekind to task on this:[9]

[I]t is impossible that the ordinals should be, as Dedekind suggests, nothing but the terms of such relations as constitute a progression. If they are to be

---

[9] Russell ([1919] 1993) eventually came around on this, sort of. In chapter 6 the mathematician is urged to adopt a version of (perhaps eliminative) structuralism, even if the philosopher cannot:

[T]he mathematician need not concern himself with the particular being or intrinsic nature of his points, lines, and planes, ... [A] "point" ... has to be something that as nearly as possible satisfies our axioms, but it does not have to be "very small" or "without parts." Whether or not it is those things is a matter of indifference, so long as it satisfies the axioms. If we can ... construct a logical structure, no matter how complicated, which will satisfy our geometrical axioms, that structure may legitimately be called a "point" ... [W]e must ... say "This object we have constructed is sufficient for the geometer; it may be one of many objects, any of which would be sufficient, but that is no concern of ours ... " This is only an illustration of the general principle that what matters in mathematics, and to a very great extent in physical science, is not the intrinsic nature of our terms, but the logical nature of their interrelations.

We may say, of two similar relations, that they have the same "structure." For mathematical purposes (though not for those of pure philosophy) the only thing of importance about a relation is the cases in which it holds, not its intrinsic nature. (pp. 59–60)

anything at all, they must be intrinsically something; they must differ from other entities as points from instants or colours from sounds … Dedekind does not show us what it is that all progressions have in common, nor give any reason for supposing it to be the ordinal numbers, except that all progressions obey the same laws as ordinals do, which would prove equally that any assigned progression is what all progressions have in common … [H]is demonstrations nowhere – not even where he comes to cardinals – involve any property distinguishing numbers from other progressions.

Frege (1903a, §§138–147) himself launched a similar attack on Dedekind's and Cantor's "free creation." Unlike Frege and Russell, Dedekind felt no need to locate the natural numbers and the real numbers among previously defined or located objects. This crucial aspect of "free creation" is shared with sui generis structuralism.

Although Dedekind was at odds with Russell and Frege here, he was still a logicist. The preface to Dedekind (1888) states that arithmetic and analysis are part of logic, claiming that the notions are "an immediate result from the laws of thought." This is a prominent Fregean theme. Dedekind added that numbers "serve as a means of apprehending more easily and more sharply the difference of things." The same idea is sounded in the 1890 letter to Keferstein:

> [My essay is] based on a prior analysis of the sequence of natural numbers just as it presents itself in experience … What are the … fundamental properties of the sequence N … ? And how should we divest these properties of their specifically arithmetic character so that they are subsumed under more general notions and under activities of the understanding without which no thinking is possible at all but with which a foundation is provided for the reliability and completeness of proofs and for the construction of consistent notions and definitions? (pp. 99–100)

The preface to Dedekind (1888) lists some of the relevant "more general notions":

> If we scrutinize what is done in counting an aggregate or number of things, we are led to consider the ability of the mind to relate things to things, to let a thing correspond to a thing, or to represent a thing by a thing, an ability without which no thinking is possible.

The general notions, then, are "object," "identity on objects," and "function from object to object." The first two of these are clearly presupposed in standard first-order logic, and the third is a staple of second-order logic.[10] These notions are sufficient to define the notion of "one-to-one mapping" and

---

[10] It is, of course, controversial whether the relevant notions from set theory are part of logic (see, for example, Quine 1986). These issues lie beyond the scope of this work.

thus "simply infinite system." Except for the infamous "existence proof" (§66) and the bit on "free creation," Dedekind's treatment of the natural numbers (and the real numbers) can be carried out in a standard second-order logic – with no non-logical terminology. In other words, once superfluous properties are jettisoned, the only remaining notions are those of logic. The relations of the structure are formal.

## 3 Set-Theoretic Structuralism

Among mathematicians, logicians, and philosophers, this is the most widely known framework for articulating mathematical structuralism. Properly speaking, the framework is called *model theory*, most of which is formalizable in axiomatic set theory. Model theory studies the relationship between formal languages and theories, on the one hand, and set-theoretic objects called structures or models, on the other, which *satisfy* the axioms of a theory, where *satisfaction* is defined inductively by the well-known methods of Tarski.[11] Roughly speaking, a model for a given theory, *T*, consists of a set or class (the domain or universe of the model), along with some relations or operations or functions on the domain, interpreting the primitive terms of the theory at hand.[12] This involves the key notion of *satisfaction* of formulas of the language based on a valuation or an assignment of relations or operations over the domain to the primitive symbols (relation- or operation-symbols) of the language of the theory being interpreted. Such an assignment or valuation determines a unique truth value (in the classical case, either truth or falsity) for each sentence of the language. When the axioms come out true, the domain, together with the relations and operations, is called a *model* of the theory. If one drops the requirement of satisfying axioms of a given theory, and considers in the abstract a domain with some designated relations, operations, and functions over the domain, one then speaks of a (set-theoretically defined) *structure* for the given language. Under this heading fall the various structures studied in abstract algebra and mappings among them (e.g., groups and group-homomorphisms), as well as the various *spaces* encountered in mainstream

---

[11] Inductive definitions of semantic relations such as *satisfaction* can be converted into explicit definitions, by well-known techniques, provided sufficiently strong existence assumptions are available in the metalanguage in which the definitions are formulated. Typically, stronger existence assumption on sets are required than available in the object language/theory.

[12] For example, a model of Dedekind-Peano arithmetic is given by specifying a set of sets representing the numbers (e.g., the finite von Neumann ordinals, with the null set as 0), along with set-theoretically defined operations, successor, addition, and multiplication (or the latter two can be explicitly defined via successor and set-membership by well-known techniques).

mathematics, e.g., function spaces of functional analysis, metric, geometric, topological, etc.

A key relation among set-theoretic structures is *isomorphism*, based on an injective and surjective map between the structures, preserving the relations and operations of the structure. The structuralist insight that, in pure mathematics, it is structure abstracting from the nature of the members of the domain that matters receives expression in the statement that mathematics is interested only in the isomorphism type of models, not in the individual models themselves.

Key properties of axiomatically presented theories are explicated in model theory, e.g., categoricity – that all models are isomorphic to one another – or various weaker relations based, e.g., on embeddings of structures or models, and so forth. The metatheorems of the soundness and completeness of first-order logic are some of the very first steps in model theory, illustrating key links between the syntax and semantics of first-order logic.

With these preliminaries understood, we can readily see how set-theoretic structuralism handles various branches of mathematics, e.g., arithmetic, analysis, abstract algebra, etc. Take the case of arithmetic. Set theory typically chooses a particular structure of sets as representing natural numbers, e.g., the von Neumann finite ordinals (finite cardinals), and then proceeds to derive the set-theoretic interpretation or translation of the Dedekind-Peano axioms from the axioms of set theory, thereby fully reducing number theory to set theory. Similarly for axioms for the real numbers, identified with, say, Dedekind cuts in the rationals, which become theorems of set theory, once the algebraic operations and the linear ordering are suitably interpreted in terms of the cuts. But to better express *structuralism*, set theory can speak of arbitrary $\mathbb{N}$-structures, defined as models of the Dedekind-Peano axioms, and then interpret sentences of number theory as saying what is satisfied in *any* such model. Similarly for real analysis ($\mathbb{R}$-structures) and complex analysis ($\mathbb{C}$-structures), and indeed any categorical mathematical theory. In the case of non-categorical theories, as in abstract algebra, e.g., group theory, set theory of course respects the multiplicity of different kinds of groups, and simply speaks of satisfaction of sentences by the relevant kind of group (i.e., model of the group axioms along with specific further conditions depending on the category of groups of interest).

Let us turn now to the criteria of evaluation of accounts of structuralism set out above in the introduction. To remind the reader, those are as follows:

(1)  What primitives are assumed and what is the background logic? Is it just first-order logic, or is second- or higher-order logic employed? If the latter,

what is the status of relations and functions as objects? What advantages and limitations are implied by these various choices?

(2) As already pointed out, the term "axioms" is ambiguous, as between "defining conditions on a type of structures of interest," on the one hand, and "basic assumptions" or "assertoric content," bearing a truth-value, on the other. As pointed out above, it is characteristic of a structuralist view of mathematics to emphasize axioms in the former sense, as defining conditions on structures of interest; and this was the sense in which, for instance, Dedekind (1888) introduced the so-called "Peano postulates" on the natural number system in his classic essay, *Was sind und was sollen die Zahlen?*; and it was axioms in this sense that Hilbert invoked in his well-known correspondence with Frege, who emphasized axioms in the assertoric sense.[13] Nevertheless, one should recognize that Frege had a point, as applied to efforts to articulate structuralism, viz. that a foundational framework requires some assertory axioms, capable of being true or false, governing especially the existence and nature of structures. (To modify a good adage: "Not by definition alone!")

(3) Are structures recognized as a special type of objects, or is there a thoroughgoing *elimination* of structures as objects? If not, what sort of objects are *structures*, and, in particular, what is a *mathematical structure*?

(4) As an especially important case of (2), what assumptions are asserted as to the mathematical existence of structures? And, especially in the case of set-theoretic structures, is their indefinite extendability recognized, or is there commitment to an absolutely maximal universe?

(5) How is our epistemic access to structures understood, and what account of reference to them can be given?

(6) Is a face-value reading of ordinary mathematical statements accommodated, or is a non-trivial translation scheme involved? For instance, are singular terms of ordinary usage treated as referring to objects on the view in question, or are they paraphrased in accordance with some translation scheme? Are quantifiers over objects of a given sort treated as such, or are they also paraphrased?

(7) Finally, how are the paradoxes associated with set-theory and other foundational frameworks (such as category theory) to be resolved? And how is Benacerraf's challenge met, to make sense of reference to numbers, for example, in the face of multiple mutually incompatible but equally good ways of construing such objects?

---

[13] Indeed, a large part of the communication problem brought out in the Hilbert-Frege correspondence is due to this basic difference in their respective understanding of "axioms." See the previous section for more on the Hilbert-Frege correspondence.

Let us take these issues up in turn with respect to set theory. As the reflective reader will point out, however, there are many set theories, and presumably the assessments will vary, depending on which version of set theory is in question. Far and away the most common choice among working set theorists would be first-order Zermelo-Fraenkel set theory (usually with the Axiom of Choice), and that shall be our focus here. But others should be mentioned that are still quite mainstream. Of special note is NBG (for von Neumann-Bernays-Gödel), with its two sorts of objects, sets and classes, where classes are the more general sort of collections, covering so-called proper classes as well as sets, where the proper classes contain members occurring at ordinally indexed levels of the cumulative hierarchy that extend arbitrarily high up, so that the class itself occurs at no such level. Furthermore, proper classes cannot themselves be members of classes. But, importantly, NBG (with Global Choice) is conservative over ZF(C), proving no theorems in the language of sets (without proper classes) not already provable in ZF(C). In contrast, Morse-Kelley set theory (MK) has proper classes, but these are governed by an impredicative comprehension scheme, allowing quantification over proper classes as well as sets in the formulas defining proper classes. This theory is virtually the same as $ZF^2$ (C), the superscript indicating that the underlying logic is second-order, with its full comprehension scheme for classes and relations. Unlike NBG, MK is not conservative over ZF(C) or NBG. (For example, MK can prove the existence of a truth-set for NBG, a set of codes of true sentences of NBG, where that notion of truth is explicitly definable in the language of MK. And then, MK can prove the statement of consistency of ZF(C), since all the axioms are true and the rules of inference preserve truth. But the statement of consistency can be given in the language of arithmetic, so certainly in the language of ZF(C).)

Let us then consider how *first-order* ZFC fares with respect to the main questions listed above, occasionally considering the alternatives recognizing proper classes when that seems especially instructive.

Re (1), the sole extra-logical primitive is the membership relation, $\in$, and the background logic is first-order predicate logic with identity. As to advantages and limitations of this choice, the main advantage concerns proof theory – the logic is demonstrably sound and complete, so that ZFC provides a tractable standard of "correct proof" for all of mathematics that can be represented in ZFC, which includes the vast bulk of mathematics as practiced.

As to disadvantages, they concern the model theory of ZFC: non-standard models abound, and there is not even an approximation to categoricity. Models can even be countable (in virtue of the downwards Löwenheim-Skolem theorem), and there can be non–well-founded models, containing infinitely descending membership chains (in virtue of the compactness theorem). This is in stark contrast with second-order ZFC (or MK), which is quasi-categorical: given any two models, one is isomorphic to an end-extension of the other.[14] From a structuralist point of view, this is a huge advantage of second-order ZFC that must be sacrificed if one chooses first-order ZFC as the framework for articulating mathematical structuralism.

Re (2), the status of axioms: The axioms of ZFC are usually taken, not just as starting points of proofs, but in the Euclidean-Fregean sense, that is, as truths about a fixed structure, the so-called cumulative hierarchy of sets. Thus, STS does not provide for a structuralist interpretation of set theory itself, but treats set theory on the cumulative-iterative conception as special. It is indeed true that first-order ZFC can represent theorizing about set-theoretic models of ZFC itself, even countable ones, in light of the downwards Löwenheim-Skolem theorem. But – reflecting the import of Gödel's second incompleteness theorem – it cannot (assuming that ZFC is consistent) prove that such models exist, much less that there is an intended maximal cumulative hierarchy.

Re (3), as the reader may surmise, structures are taken to be sets with set-theoretically defined relations and functions over them, and so STS is certainly not *eliminative* with regard to structures as objects. There is, however, a *de facto* elimination of structures other than those representable as sets. For example, there is no structure recognized of *sui generis* natural numbers or real numbers, etc., but only set-theoretic models of the associated theories, deploying conventionally chosen identifications of the individual objects, as in the case of, say, the von Neumann finite ordinals for natural-number theory, or Dedekind cuts in the rationals for real analysis, etc. Thus, as in the case of the status of axioms, set theory is given special treatment.

As to what counts as a *mathematical* structure, a natural criterion would be any set-theoretically definable structure. But then, what about structures modeling extra- or applied-mathematical theories, e.g., for physics, chemistry, economics, etc.? It is a conventional matter whether to call such structures "mathematical" or perhaps "applied mathematical," recognizing the special, non–set-theoretic vocabulary associated with their models.

Turning to (4), the existence of mathematical structures, a virtue of STS is its clarity on this. Structures are recognized if they can be proved to exist in ZFC. Under this heading fall all the structures encountered in mainstream mathematics, including algebraic, geometric, topological, functional analytic, and so forth. What about structures for extraordinary mathematics, e.g., for set theory

---

[14] This was first established by Zermelo in his great paper of 1930, where he proved that models can differ along only two parameters (the cardinality of the urelement basis and the ordinal height), and where he also established that the ordinal height of any model is a strongly inaccessible cardinal.

itself? Here significant limitations arise, as implied by the work of Gödel, Tarski, and others: on pain of contradiction, ZFC cannot prove the existence of models of ZFC, as that would amount to proving the consistency of ZFC in ZFC, which is ruled out by Gödel's second incompleteness theorem (as applied to this case, if ZFC is indeed consistent, then it cannot prove the standard statement of its own consistency). If, however, so-called large cardinals are recognized – e.g., by adding an axiom stating that there exist strongly inaccessible cardinals, then this implies the existence of models of ZFC *simpliciter*, without the extra axiom. As the study of large cardinals shows, this is just the first step in a towering hierarchy of large cardinal axioms, exhibiting a pattern of larger and larger cardinals demonstrating the existence of smaller ones and also cumulative hierarchies as models of theories with smaller large cardinal axioms.

This naturally brings us to the question about indefinite extendability of universes for set theory. Despite the possibility of enriching ZFC with axioms for progressively larger and larger cardinals, one encounters significant limitations, reflecting the fact that STS perforce makes a massive exception in the case of set theory itself: that is, at some level, *it does not treat set theory structurally* in that its axioms are not understood as defining conditions on structures of interest but rather as true assertions about a fixed, maximal background, *the* cumulative hierarchy of sets (or even "the real world of sets"), and this sets an upper limit to the extendability of models of set theory. In short, whereas set-sized models of set theory go on and on with stronger and stronger axioms of infinity, the fixed background universe of "all sets" is treated as absolutely maximal, a glaring exception to the principle of indefinite extendability. Conceptually this is problematic: for given any totality of sets whatever, what is to prevent the mathematician from taking that totality as a higher type of collection subject to operations strictly analogous to the familiar set-theoretic ones, e.g., of taking singletons, or power-collections, as applied to the putatively maximal model or universe itself? Thus, there is an arbitrariness in the cases of NBG and MK, which explicitly recognize the proper class of all sets or all ordinals, with their not accommodating still higher type totalities (sometimes called "supersets"); whereas retreating to first-order ZFC only postpones the problems associated with recognizing the putatively maximal universe of sets. In practice, one implicitly recognizes "all the sets," at least in a plural sense, even if one does not officially embrace proper classes.

Turning to the fifth item on our list, epistemic access and reference to structures, in STS this is covered by reference to sets along with distinguished relations and operations on them. Typically, one begins with the null set (an axiom provides that there is such a thing); then one considers its singleton, then the pair of the null set and its singleton, and so on, iterating the power set operation through all finite levels of a cumulative hierarchy. Next, an explicit Axiom of Infinity guarantees that there is a set of all the finite von Neumann ordinals. Then the hierarchy continues, by iteration of the power set operation, taking unions at limit levels, beyond which more sets are assumed to arise, as guaranteed by the powerful Axiom of Replacement, which implies that the cumulative hierarchy continues beyond any level whose predecessors are in one-to-one correspondence with any existing set. (The hierarchy is inevitably "taller" than it is "wide.") Thus, the power and flexibility of the set-theoretic machinery allows one to identify all the structures encountered in ordinary mathematics with sets and relations occurring in this massive cumulative hierarchy.

Whether, when one takes power sets and unions in "generating" the hierarchy, they are really full, omitting no sets along the way, is an esoteric question that mathematicians would never ask or worry about. But this only shows that something is amiss with a straightforward, face-value reading of set-theoretic discourse. Perhaps, for example, mathematicians only mean to be describing mathematical possibilities, not an actual set-theoretic universe. In any case, philosophically, such questions do seem intelligible, and they seem hard to answer so long as one takes seriously the idea that set theory really is describing a pre-existing, actual hierarchy of definite objects. This realist (almost physical) picture seems beset with such questions because they are not blocked by stipulations. The objects-Platonist stance prevents one from saying, simply, that we are interested in studying "full" power sets, or a hierarchy or hierarchies of sets obeying the Replacement Axiom: it always seems pertinent to ask whether there really *are* such full power sets, or whether some of them have "gaps." Or how do we know that, in reality, the cumulative hierarchy itself cannot be put into one-to-one correspondence with any item occurring at a given level? Again, such questions take us away from mathematics per se, but they persist in annoying the philosopher-logician who is seeking to articulate a fully justifiable structuralist account.

Concerning item (6), whether a face-value reading of mathematics is accommodated, this depends on whether the mathematics is set theory itself, or (as is typical) a theory of mainstream, ordinary mathematics, e.g., theories of numbers or the various spaces usually investigated. In the case of set theory itself, a face-value reading is standardly adopted. But all other theories have to be translated into the language of sets. Ordinary mathematical theories thus do not stand on their own. Moreover, there is a well-known multiplicity of, for instance, identifications of numbers as sets, giving rise to the well-known problems highlighted by Benacerraf (who argued that numbers cannot really *be* sets, for then they would have to be particular ones, but any choice would arbitrarily favor one set of identifications over all other equally legitimate ones). (We return to this below.) Indeed, a common criticism of STS is that its representations of ordinary mathematics are highly artificial and cumbersome, very far removed from ordinary practice.

In reply, the set theorist can point out that the aim is not to capture ordinary usage but rather to provide a rational reconstruction including providing a uniform standard of mathematical existence and, equally important, a uniform standard of correct mathematical proof. It is by no means intended that set-theoretic translations and representations be employed in ordinary mathematical practice.

Last on our list there is the question of how STS deals with the well-known paradoxes of set theory, especially the Burali-Forti, Russell's, and Cantor's paradoxes, respectively associated with "the largest ordinal," "the set of exactly all non–self-membered sets," and "the set of all sets." To be clear, the main systems of set theory do successfully block these paradoxes from arising, although the details differ depending on the system. But, as we will explain, that is not the end of the matter, as certain sacrifices must be made to achieve the standard set-theoretic resolutions. Take, for example, first-order ZFC. There simply is no ultimate or largest ordinal, as is proved by *reductio*: if there were, call it $\Omega$, then it would have a successor, $\Omega + 1$; but then since $\Omega$ dominates all ordinals, it would have to dominate also $\Omega + 1$, which is a contradiction. For purely mathematical purposes, then, first-order ZFC need not worry about Burali-Forti, since nothing in the role of $\Omega$ is recognized. But, at least informally, there still is the well-order relation $\prec$ on the ordinals: although it is not recognized as a set, informally one recognizes it, in the case of the von Neumann ordinals, as the membership relation itself restricted to those ordinals. But then there is no ordinal representative of that relation, as described plurally (as "all the pairs $\langle x,y \rangle$ such that $x \prec y$"). At the least, there is a lack of correspondence between what one can recognize on reflection and what is officially countenanced in one's theorizing. What is the situation if, instead of ZFC, one moves to NBG or to MK? Then one does recognize the well-ordering relation among "all the ordinals," and then one also recognizes $\Omega$ as the ordinal representative of the well-ordering. But paradox is blocked because $\Omega$ is a proper class and so not a member of anything, by stipulation. So, although $\Omega$ represents the well-order relation on the ordinals, it does not behave as an ordinal, lacking any successor. Nevertheless, the coherence of adding supersets to continue the cumulative hierarchy of sets points to the arbitrariness of resting with NBG or MK. Coherent extensions of any hierarchy of sets are surely mathematical possibilities, so the resolution via proper classes accords ill with the freedom of the mathematical mind to entertain ever more comprehensive structures of set-like objects.

Similar remarks apply to the standard resolutions of the Russell and Cantor paradoxes. While these are clearly blocked, either by the axiom schema of restricted comprehension or Separation (*Aussonderung*) in the case of ZFC, or by the limitations on collections imposed by NBG and MK,[15] there is a significant price, viz. that set theory cannot claim to be a fully general theory of collections and membership among them. Not every plurality of sets can be represented as a set, on pain of generating paradox. Put otherwise, by virtue of extendability, one can always continue to consider richer and richer structures of set-like objects, but those cannot be identified as the officially recognized sets and proper classes. It is a challenge to improve on the standard set theories, but we shall see that some alternatives have been proposed in efforts to meet the challenge.

Finally, what should the set theorist say about Benacerraf's problem about multiple, equally good translations or models of ordinary mathematical theories? Of course, such multiplicity should be acknowledged; and it should be conceded that there is no unique, correct set-theoretic representation of, say, the natural numbers, or the real numbers, etc. But there simply is no claim that numbers "really are" these sets or those, etc. Indeed, strictly speaking, the predicate "natural number" is eliminated in favor of, e.g., "finite von Neumann ordinal." It can cheerfully be granted that other choices are equally good.

Thus, to summarize, STS, despite its clarity and definiteness, and its clear success in representing the wide variety of structures arising in ordinary mathematics, is not without its problems. To review, the main ones are these three:

(i) As we have seen, set theory itself is not treated structurally, despite the fact that there are many different set theories differing on various axioms, yet representing legitimate mathematical investigations. Thus, as already seen, set theory is deprived of one of the most natural ways of justifying its axioms, viz. by stipulative appeal to the mathematicians' interest in investigating them. Consider, for example, the matter of well-foundedness (enforced by the Axiom of Regularity). STS, it appears, is called upon to defend this as a true description of a pre-existing set-theoretic

---

[15] In ZFC, the axiom scheme of Separation allows there to be a set of only all non–self-membered sets that already belong to a given set, thereby blocking Russell's paradox. And in NBG or MK, there is only a proper class of all sets that are non–self-membered, and that class cannot belong to anything.

reality, despite the coherence and legitimacy of mathematical investigations into non–well-founded set theory. This is a burden that a version of structuralism should not have to bear. It would be perfectly defensible to cite our interest in studying well-founded structures as sufficient justification of the Axiom of Regularity, without having to say that set theories without Regularity are somehow a false description of set-theoretic reality. Or consider the problem of justifying the Axiom of Replacement. It does now seem that the "iterative conception of set" (as articulated by Boolos 1971) can be enriched to provide a derivation of Replacement (despite Boolos's doubts).[16] But in any case, one can *stipulate* that we are interested in exploring structures so large that their extent cannot be "measured" by any of the elements making them up. This should not involve any implicit claim that set-theoretic structures that are small in this sense are not a fit subject of mathematical investigation, or that they are lacking due to failure to "correspond with set-theoretic reality." Thus, pluralism clearly seems indicated, but STS as usually formulated seems committed against it.

(ii)  STS seems to have to confront esoteric and intractable questions concerning its axioms, taken in their literal sense. How, for instance, can we be sure that there really is such a thing as the null set? Should the viability of classical mathematics rest on such an assumption? Would it not be better to regard this as a convenient fiction? Or what about the assumption of full power sets? What assurance do we have that set-theoretic reality respects this, that there really are not any gaps? Would it not be better somehow to free mathematics from such an assumption, somehow to block or bypass such a question entirely?

(iii)  Finally, STS seems unfortunately committed to the existence of a maximal universe, whereas this contravenes the natural mathematical principle that any totality whatever can be transcended by a more comprehensive one. Indeed, this is so in virtue of set-theoretic operations themselves, e.g., taking the singleton of any given totality, or taking the power set, etc. As Zermelo (1930) put it, this capacity to go beyond any given totality is rooted in the creativity of mathematical thought, and somehow our foundations of mathematics should respect and reflect that.

---

[16]  See, for example, Hellman (forthcoming) for ways of carrying this out. But note that his proposals are acceptable only on a "height-potentialist" view of set-*cum*-stage theory, according to which any possible stages are possibly followed by additional ones, but not on a "height-actualist" view, recognizing "all stages." Cf. also Linnebo (2013) for another height-potentialist derivation of Replacement.

Thus, despite the clear successes of model theory in providing a structuralist account of ordinary, mainstream mathematics, problems such as these have motivated some mathematicians, logicians, and philosophers to seek other ways of articulating a structuralist framework. We turn to these in the succeeding sections.

## 4 Category Theory as a Framework for Mathematical Structuralism

As in the case of set theory, category theory (CT) arose, not as a foundational framework at first, but in order to solve problems within mathematics itself. In the case of category theory, it was first developed by Eilenberg and Mac Lane in the 1940s, in response to problems in algebraic topology and algebraic geometry. Only later, with the formulation of the notion of a topos – generalizing the concept of a cumulative hierarchy of sets, but deploying the language of categories and functors, and allowing the associated logic to be intuitionistic (with classical logic governing only special cases) – did Lawvere (1964, 1966), Mac Lane (1986), and others propose topos theory as providing a foundational framework for mathematics generally, claiming that this stood autonomously, independent of set theory. And more recently still, Awodey (1996) has explicitly suggested that category and topos theory provide an independent and natural framework for realizing mathematical structuralism, one that philosophers of mathematics should study and develop.

Awodey's idea is that category theory, in virtue of its apparatus of "objects" and "arrows" – where the objects are typically whole structures and the arrows are typically structure-preserving mappings or "morphisms" between objects – is in an especially good position to articulate and illuminate mathematical structure(s). To appreciate how this arises, consider what a *category* is: an algebraic structure defined by a list of conditions stipulating that arrows between pairs of objects can be composed with other arrows (when the codomain of one arrow serves as the domain of another), the composition being associative and respecting the identity arrows (or morphisms) assigned to each of the objects. A *category* then is *anything* that jointly satisfies these defining conditions. And a mapping between two such categories qualifies as a functor if it preserves the compositional structures of their arrows or morphisms. What is remarkable about this apparatus is that the vast bulk of structures arising in mathematics can be captured in the spare language of " arrows only," without "looking inside" the objects (which, recall, are typically whole structures or spaces, e.g., algebraic, metric, topological, etc.).

As an important illustration, consider the very beginnings of how category theory describes hierarchies of sets. It abstracts from the membership relation, availing itself instead with an externally defined correlate thereof, deploying "arrows only." This arises by introducing the notion and positing of a "terminal object," $t$, one such that for any other object, $o$, there is but a unique arrow $a$ mapping $o$ to $t$. Then there will be a one-to-one correspondence between the set-theoretic elements $e$ of a set $s$ and the arrows from a terminal $t$ to $s$. This provides the bare beginnings of a kind of reduction of membership-based set theory to a category or topos of sets.[17]

An important advantage of the CT approach to mathematical structure is that it abstracts from the means by which a type of space or structure may have been introduced. Topological structure, for example, is captured in terms of the continuous maps between and among whole spaces, without reference to "open sets," "limit points," closure operations, and so on. Furthermore, CT's treatment of spaces and structures illuminates how the objects involved need be distinguished only "up to isomorphism," one of the key motivating ideas of mathematical structuralism.

When it comes, however, to assessing the proposal of category theorists that CT serve as a framework for mathematical structuralism, we confront the situation that we lack a generally agreed-upon system of axioms as assertions constituting the framework. After all, the axioms mentioned above defining what a category is, are just defining conditions on a type of algebraic structures. They figure as starting points in the practice of category theory *qua* mathematics proper. But this must be distinguished from category and topos theory as a foundational framework, that is, if CT structuralism is to be deployed the way STS and SGS are in the interests of foundations of mathematics. In other words, such a framework requires both kinds of "axioms," some defining its key notions of *category* and *functor*, and others consisting of positive assertions concerning the existence and inter-relations of categories and topoi. The practice described by Awodey, et al., refers to the former algebraic "axioms" but is silent on the latter. And Mac Lane (1986) has proposed that a topos meeting certain conditions guaranteeing extensional discrimination of functions (a "well-pointed" topos) could constitute a universe of discourse for mathematics on a par with set theory's cumulative hierarchy. Similarly, Bell (1986) has proposed that other topoi can also serve as universes of discourse for the development of mathematics, placing topos theory on a par with set theory. For example, such topoi have machinery corresponding to CT conditions defining Cartesian products, functional exponentiation, and other key operations giving rise to an internal logic (through what is called "subobject

---

[17] Of course, a great deal more machinery is involved. The interested reader can consult, e.g., Goldblatt (2006).

classification"), analogous to and generalizing set-theoretic operations. Further, a topos must be endowed with a natural numbers object if the reconstruction of the number systems is to be provided. But once again, these are just defining conditions on a special type of hierarchical structures. They are clearly central to the practice of category theory as mathematics proper; but they do not by themselves address the issues we have listed as essential to a structuralist framework, *as a chapter in metamathematics*.

Closely related to these points is the critique Feferman (1977) mounted against Mac Lane's claims that CT provides a viable alternative foundation to set theory *vis-à-vis* ordinary mathematics. In essence, Feferman emphasized the fact that CT in practice relies heavily on the notions of *collection* and *operation,* both in describing what a category or a topos is, and in describing how categories are interrelated by homomorphisms or functors. Furthermore, he argued, a foundational framework must provide a systematic account of these notions (*collection* and *operation*); this is clearly something that set theory does, but CT does not, at least not with its algebraic descriptions of "category," "functor," etc., and not through the mathematical practice of proving theorems about categories and functorial relations. Feferman left open the prospect that some alternative framework to set theory could provide the requisite account of "collection" and "operation," that is, he was not insisting that CT depends on set theory per se, but simply that, as it stands, it is inadequate as a foundational framework.

Now there is a possible response to all this that needs to be considered, namely the claim that CT investigates the behavior of families of functions (via its arrows, related by composition as exhibited in its characteristic diagrams), and that this is not in principle different from what set theory does as a study of collections and operations. Indeed, the argument goes, the notion of "function" is one that mathematics uses effectively all the time; indeed, set theory itself can be presented in terms of functions (as characteristic functions of collections, as in von Neumann's [1925] axiomatization of set theory). Thus, CT could be seen as an alternative way of theorizing about functions, arguably in many instances more closely tracking mathematical practice than via set theory. What should we make of this sort of argument?

We see two problems with it. First, it depends on a privileged interpretation of the arrows of a category, whereas there are contexts where the arrows are intended to be interpreted differently, e.g., as proofs of propositions in applications of CT to formal systems of proof. Thus, reading CT as a theory of functions is not always faithful to the practice at hand.

Second, however, one still needs assertory axioms on the existence of functions, even if we confine our interpretation of "arrows" in the manner suggested. For example, what does CT say about axioms of Infinity and Choice, e.g., the existence of global choice functions? What about well-foundedness? Or the (arrow-theoretic) content of the set-theoretic Replacement axiom(s)? (All these questions can be framed in terms of functions, hence in terms of arrows on the present proposal. But assertory existence axioms are needed to provide cogent answers.) And there are questions specific to category theory that need to be addressed in assertory axioms, especially pertaining to the nature and scope of categories. For example, is there really a category of *all* categories? If so, how is an analog of Russell's paradox to be avoided? (We return to this question below.)

Thus, if we turn to the questions/issues on which we have been comparing the various approaches to structuralism, we find ourselves in the awkward position that, with the exception of the third question, there seem to be no definite answers provided due to the underspecification of assertory axioms at the foundational level. Consider (1), the choice of primitives and background logic. Here we are not asking about the definition of "category" or "functor," but of a substantive metatheory pertaining to these notions. We cannot even say that first-order logic is sufficient (and below we shall raise doubts that it is). Regarding question (2), specifying the assertory axioms of the framework, the absence of these has been our main complaint thus far. It is true, however, that we can address the third question on whether structures are eliminated as objects. Pretty clearly the answer is no, despite the absence of existence axioms on categories and functors, for clearly categories and topoi are recognized as structures in terms of which the vast bulk of mathematics is to be understood. But what about the other subquestions under (3): the nature of structures, and what distinguishes *mathematical* structures from others? As objects within categories, structures are treated as "point-like," with arrows among them doing all the work of delineating their mathematical properties. As remarkable as this is, one may doubt that it would be intelligible without prior background experience with concepts based on set-membership, or experience with (symbols representing) numbers as objects.

Concerning point (4), the existence of structures, our main complaint leveled against CT structuralism has been that this is underspecified. It should be pointed out, however, that texts on category theory sometimes defer such questions to a background set theory. (For example, Bell suggests that CT and topos theory be developed in the framework of NBG set theory.) But of course this is not an option for the CT structuralist (such as Mac Lane) who wants the framework to stand independently of set theory. With regard to the second part of (4), the extendability of structures, informal remarks of Mac Lane, Bell, and others suggest that they do subscribe to a principle of indefinite extendability, and they see CT and topos theory as an improvement over set-theoretic foundations on this score. However, without formalization via existence postulates, it remains unclear how this is to be implemented.

When it comes to point (5), reference and epistemic access to structures, too little has been said for us to comment on.

Thus far, our overall complaint has been that the key ingredients in articulating a framework for mathematical structuralism (summed up in our five questions) have been underspecified, or else are to be provided by a background set theory, which for present purposes, is less than satisfactory, since an autonomous framework independent of set theory is sought. In response to all this, however, McLarty (2004) has called attention to some axiom systems that can perhaps fill the void, consisting as they do in axioms as assertions taken to be true. Two of those systems are relevant here and call for comment, viz. the system known as ETCS – elementary theory of the category of sets – and a system known as CCAF – category of categories as a foundation. (These are due to Lawvere 1964 for ETCS, and 1966 for CCAF.) Let us now consider these and how they compare to set-theoretic foundations.[18]

To begin, ETCS is a theory of sets described entirely in terms of objects and arrows. Intuitively, one thinks of a well-founded universe of objects arranged in a hierarchical manner in levels indexed by objects playing the roles of ordinal numbers. Thus there are powerful existence axioms of Infinity, Power Sets, Replacement, and Choice, much as in standard, element-based set theory. (It is something of a *tour de force* to describe all this entirely in terms of arrows. Even reference to objects can be eliminated in favor of identity arrows, as constrained by the axioms defining "category.") Thus, these axioms specify a special, highly structured category called a well-pointed topos of sets, suitable for expressing and deriving virtually all of classical mathematics.

From a structuralist point of view, it is certainly noteworthy that this topos of sets treats its objects as defined "only up to isomorphism," and that this provides structuralist insights by focussing on the interrelations among the objects of this category, in accordance with the demands of "arrows only" descriptions. Clearly, we've come a long way from just the elementary conditions defining what a category is. What further should be said about this

---

[18] In addition to the summary that follows, the interested reader may also consult Hellman (2003) and follow-up exchanges, including replies by Awodey (2004) and McLarty (2004), and my reply to these, Hellman (2006).

It should also be mentioned that Homotopy Type Theory has also been proposed as a foundational framework with assertory axioms. Assessing this, however, exceeds the scope of this volume.

categorical set theory, ETCS, as providing a structuralist foundation of mathematics?

Of the seven groups of questions we've been putting to the various versions of mathematical structuralism, the fourth and fifth stand out as posing the greatest challenge. Just as in the case of element-based axioms systems, like ZFC, there is either implicit or explicit commitment to an absolutely maximal universe of discourse for mathematics, counter to the Zermelo-Putnam extendability principle, concordance with which we've been taking as a *desideratum* on any structuralist account. This is all the more significant in that, as a foundational framework, ETCS is supposed to incorporate the development of all known mathematics. Furthermore, commitment to an absolutely maximal background universe runs directly counter to the professed views of leading category theorists, like Mac Lane, who has expressed this as follows:

> There are no upper limits. It is useful to consider the "universe" of all sets, or the category *Cat* of all small categories, or the category CAT of all big categories.... [but] after each careful delineation, bigger totalities appear. No set theory and no category theory can encompass them all – and they are needed to grasp what Mathematics does. (1986, p. 390)

Concerning point (5), epistemic access and reference to mathematical structures, there is the question whether ETCS is epistemically autonomous, or whether understanding these axioms depends on a prior grasp of element-based set theory.[19] It seems to us that the latter has deeper roots in ordinary everyday practice regarding collections, and that "arrows only" methods of articulating set-theoretic interrelations come later, after quite a lot of experience with sophisticated, abstract structures like vector spaces and other function spaces. Thus, it is a serious question, especially when it comes to issues of epistemic dependence, whether CT structuralism can stand autonomously in relation to element-based set theory.

What about point (6), on allowing for a face-value reading of ordinary mathematical discourse? Well, here the departure from ordinary discourse is even greater than it is with ZFC set theory, as "arrows only" formulations are that much further removed from ordinary discourse than the language of element-based sets and classes. (This is already suggested by our remarks concerning point (5) on epistemic access and reference to structures.) As it turns out, however, most foundational frameworks depart substantially from ordinary discourse, and one can regard this as a price that has to be paid in order to meet other more pressing goals. (The main exception to this is SGS, at least with regard to number systems and analysis.)

---

[19] The interested reader should also consult Linnebo and Pettigrew (2011).

There is, however, a sense in which CT structuralism comes close to a face-value reading, at least of the number systems, in the context of ETCS. There, one has as a postulate or theorem that a natural numbers object N exists. From this, it follows that there is an initial "element" of N ("element" in the CT sense of arrow from a terminal object to N), and also a unique "successor" of this, and so on. This justifies introducing numerals, behaving like names but also like free variables, subject to generalization. The upshot is a reading of the number systems quite similar to that available to the MS framework, commented on below.[20]

Regarding point (7), about resolving paradoxes, ETCS can mimic the standard set-theoretic resolutions, but it does not appear to improve upon the latter in connection with, e.g., ordinal representability of well-orderings, or with set representability of arbitrary collections, etc.

What should be said, on behalf of ETCS, regarding Benacerraf's challenge based on equally good but conflicting ways of identifying numbers and other items of ordinary mathematics? Here, the resolution is similar to the story that STS, MST, and MS can tell. As just mentioned in connection with point (6), the axioms guarantee, for instance, the mathematical existence of a "natural numbers (NN) structure," hence a sequence with an initial item, call it "0," *succeeded* (in the sense of the arrow of the NN structure) by a unique further item, call it "1," which is succeeded by a unique "next" item, call it "2," etc. But these numerals are being treated as variables, dependent on the given NN object, rather than genuine proper names having absolute referential significance. They arise via existential instantiation; following proofs of theorems of number theory, they are subject to universal generalizations holding of arbitrary NN objects. At no point is it ever claimed that "these are really the natural numbers." In this sense, Benacerraf's problem simply does not arise.

Finally, regarding ETCS, let us point out that another problem we found confronting set-theoretic structuralism also confronts the present theory, viz. what we called the problem of "missing sets." It seems that any set theory, regardless of its primitives, confronts this embarrassment – e.g., that real-world power sets might not be "full" – so long as it is taken as describing an actually existent domain of objects (whether traditionally conceived as collections, or conceived as functions, or schematically as arrows).

Thus, despite the promise of ETCS, we are motivated to inquire about other assertory axiom systems based on the concepts of category theory.

---

[20] Both CT and MS perspectives can avail themselves of what Pettigrew (2008) calls "dedicated free variables," subject to rules of existential instantiation and universal generalization. This narrows the gap considerably between these approaches and so-called "face-value" interpretations. A similar story could be told also about STS and MSTS.

Turning now to a category of categories, efforts towards axiomatization are at least grabbing the bull by the horns, laying down explicit assertory axioms on the mathematical existence of categories, and providing a unified framework for a large body of informal work on categories and topoi (and hence mathematics, generally). Three questions demand our attention:

(1)  What concepts are presupposed in such an axiomatization?
(2)  Are these such as to sustain the autonomy of CT vis-à-vis set theory or related background, or do they reveal a (possibly hidden) dependence thereon?
(3)  What is the scope of such a (meta)theory, and in particular, what are the prospects for self-applicability and the idea of "the category of (absolutely) all categories"?

On (1) and (2), it is clear that these axioms – framed as a first-order theory (as in McLarty 1991) – are not employing the CT primitives ("object," "morphism," "domain," "codomain," "composition"), schematically, as in the algebraic defining conditions, but with intended meanings presumably supporting the at least plausible truth of the axioms. The objects are categories, the morphisms are functors between categories, etc. Commenting on this, Bell and I have written:

> Primitives such as "category" and "functor" must be taken as having definite, understood meanings, yet they are in practice treated algebraically or structurally, which leads one to consider *interpretations* of such axiom systems, i. e. their semantics. But such semantics, as of first-order theories generally, rests on the set concept: a model of a first-order theory is, after all, a set. The foundational status of first-order axiomatizations of the [better: a] metacategory of categories is thus still somewhat unclear. (Hellman and Bell 2006, pp. 74–75)

In other words, when we speak of the "objects" and "arrows" of a metacategory of categories as categories and functors, respectively, what we really mean is "structures [or at least "interrelated things"] *satisfying* the algebraic axioms of CT," i.e., we are using "satisfaction," which is normally understood set-theoretically. That is not to say that there are no alternative ways of understanding "satisfaction"; second-order logic or a surrogate such as the combination of mereology and (monadic) plural quantification of modal-structuralism would also suffice. But clearly there is some dependence on a background that explicates "satisfaction" of sentences by structures, and this background is not category theory itself, either as a schematic system of definitions or as a substantive theory of a metacategory of categories. But this need for a background theory explicating "satisfaction" was precisely the conclusion we came to in our 2003 paper, reinforcing the well-known critique of Feferman (1977), which exposed a reliance on general notions of "collection" and "operation." It was precisely to demonstrate that this in itself does not leave CT structuralism dependent on a background set theory that I proffered a membership-free theory of large domains as an alternative. Although the reaction, "Thanks, but no thanks!", frankly did not entirely surprise me, it will also not be surprising if a perception of dependence on a background set theory persists.

As to the third question of scope, I think it is salutary that McLarty calls his system "*a* (meta) category of categories," rather than "*the* category of categories," which flies in the face of general extendability. No structuralist framework should pretend to "all-embracing completeness," in Zermelo's (1930) apt phrase. And we certainly had better avoid such things as "the category of exactly the non–self-applicable categories," where a category is "self-applicable" iff it occurs as one of its own objects. (For this leads straight to an analog of Russell's paradox: if there were such a category, it would be self-applicable if and only if it were not.)[21]

To conclude this section, it seems a fair assessment to say that, while axioms for a (meta)category of categories do make some progress toward providing answers to some of our seven questions put to the various versions of mathematical structuralism, we are left still well short of satisfactory, full answers, even to the first four questions.

## 5 Structures as *Sui Generis* Structuralism

As noted in the historical introduction, Section 3 above, a good precursor to this approach can be found in the writings of Richard Dedekind, especially his monograph, *Was sind und was sollen die Zahlen* (1888), together with some of his correspondence with Heinrich Weber. In the monograph, Dedekind introduces the crucial notion of a *simply infinite system*, a set of objects along with an operation or one-to-one function defined on the set, explicitly obeying four conditions, including a second-order or set-theoretic statement of *mathematical induction*. These conditions were taken over by Peano, and are now known as the "Peano postulates" (although Peano explicitly acknowledges Dedekind as his source). But crucially, Dedekind does not frame them as postulates or

---

[21] For a good relative consistency result on axioms for a category of categories (consistency relative to axioms for a well-pointed topos including separation and a natural-numbers object), see McLarty (1991). Such a topos is close to a model of Zermelo set theory, so it is reasonable to expect that an analog of Russell's paradox cannot be derived.

Also, McLarty's axioms do not assert the existence of any (meta)category of categories; rather they posit the existence of many categories and functors between them, without positing any category of all of those. In this framework, which could also be formulated with the language of plural variables, there seems no way to formulate a Russell-type paradox.

axioms, but rather as *defining conditions* on a *type of structure* of interest. These function as "axioms" in the algebraic sense associated with Hilbert and modern structuralism, not as assertions taken to be true (axioms in the sense associated with Euclid and Frege). In a key passage, Dedekind (1888, §73) wrote:

> If in the consideration of a simply infinite system $N$ set in order by a transformation $\varphi$, we entirely neglect the special character of the elements, simply retaining their distinguishability and taking into account only the relations to one another in which they are placed by the order-setting transformation $\varphi$, then are these numbers called *natural numbers* or *ordinal numbers* or simply *numbers* ... With reference to this freeing the elements from every other content (abstraction) we are justified in calling numbers a free creation of the human mind. The relations or laws which are derived entirely from the [defining] conditions are always the same in all ordered simply infinite systems, whatever names may be given to the individual elements.

As pointed out in Section 3 above, this passage can be understood as opting for an eliminative structuralism, associated with Hilbert (see also Shapiro 1997). However, it could also be taken as positing a particular, privileged system or structure, obtained from all particular realizations via a kind of abstraction, a system of universals – positions or places in a special abstract structure. Such a philosophy has been proposed by contemporary philosophers of mathematics, Michael Resnik and Shapiro, under the latter's rubric, *ante rem* structure.[22] Here, we dub the view *sui generis* structuralism, abbreviated SGS.

Under this latter interpretation, the very distinction between Hilbertian axioms and Euclidean-Fregean axioms collapses, or perhaps it would be more accurate to say that Dedekind's defining conditions also serve as basic truths about the particular abstract numbers he introduced. Again, most of Dedekind's published writings can reasonably be given either interpretation, although there is textual evidence in *Was sind* (1888) for interpreting Dedekind as an *ante rem* structuralist. For example, when it comes to proving in effect the categoricity of the Dedekind-Peano "axioms," Dedekind (1888, §132) writes that "all simply infinite systems are isomorphic to $(N, \varphi)$, and therefore to one another." Clearly here "$N$" and "$\varphi$" are behaving as proper names, not variables.

In correspondence with Weber, Dedekind clarified his position. In discussing his theory of real numbers, he concedes that, as far as pure mathematics is concerned, identifying the irrationals with the (open) cuts in the rational

---

[22] Shapiro uses this term to contrast with *in re* structures: the latter comprise *places* "as offices," filled by particulars, e.g., sets or regions of a space, etc., whereas the former comprise *places* as *sui generis* objects in their own right.

numbers is harmless, but he prefers to introduce "something new, which the mind creates," distinct from the cuts themselves (Dedekind 1932, vol. 3, pp. 489–490). At least with hindsight, this is clearly opting for a *sui generis* or *ante rem* option. It is thus natural to interpret the talk of "free creation" in the above passage from Dedekind (1888) in similar terms.

Inspired by Dedekind's ideas about structures obtained via abstraction from the particular identities of systems of objects meeting the structural defining conditions, Shapiro (1997, pp. 93–97) develops a formal theory, *ante rem* (a.r.) structure theory, in the setting of second-order logic. First-order variables range over structures and "places" within structures. Intuitively, "systems" are classes of places, together with certain relations and functions thereon. Systems are values of the second-order variables.

The second-order background logic contains all instances of second-order logical comprehension of the form

$$\exists R \ \forall x_1, \ldots, \forall x_n (R(x_1, \ldots, x_n) \leftrightarrow \Phi(x_1, \ldots, x_n)),$$

where $\Phi$ is a second-order formula lacking free "$R$," but possibly containing both first- and second-order bound variables. This, of course, is full, impredicative second-order comprehension for classes (with $n = 1$) and relations (when $n > 1$). Added to this background are axioms on *ante rem* structures modeled on the axioms of ZFC set theory (with second-order Separation and Replacement), including the existence of infinite structures, the existence of power-structures, a form of second-order Replacement Axiom, and a scheme of second-order Reflection (implying Infinity and Replacement, along with many small large cardinals).[23] Crucially there is also an Axiom of Coherence, using a new primitive predicate, *coherent*, the a.r. structuralist analogue of joint satisfiability in model-theory. This axiom asserts that any class of second-order sentences that is coherent characterizes at least one *ante rem* structure. Of course, all the axiom systems appealed to in ordinary mathematics are assumed to be (individually) coherent.

Thus, *ante rem* structure theory is certainly not eliminativist with respect to structures as objects. So long as a system of axioms is coherent and categorical, then there is a unique a.r. structure satisfying those axioms. To anticipate, regarding our criterion (8) of evaluation, we note that this provides a ready answer to Benacerraf's puzzle. While there are indeed many ways of *modeling* numbers as sets (be they natural, rational, real, or complex numbers, etc.), only one of these models is the distinctive a.r. structure of numbers themselves.

---

[23] In Shapiro's 1997 book, an axiom of choice is not presented, but it can be added as a friendly amendment, or else formulated in the background second-order logic.

Officially, number words refer to the places or positions of the relevant *ante rem* structure. *Ante rem* structures are thought of as abstract types exemplifying "what all particular realizations of the axioms have in common," where some of the latter are already systems of abstract particulars (such as sets). From a nominalist perspective, the *ante rem* account thus qualifies as "hyper-Platonist." We will return to this answer to Benacerraf's challenge below.

Turning now to our points of comparison and assessment, regarding (1) and (2), we have already reviewed the primitives of SGS, at least in connection with Shapiro's *ante rem* structure theory. The background logic is second-order, with the full, impredicative comprehension scheme, which provides a theory of relations and functions. The first-order variables range over *ante rem structures*, and also over "*places*" or "*positions*" in a.r. structures. Regarding extra-logical axioms, we have seen that, in Shapiro's system, they function *both* as "algebraic," defining conditions of types of structures, *and* as basic assertions giving specific information about *ante rem* structures and operations on them.

Concerning point (3), on the status of structures as objects, we have already pointed out that SGS is non-eliminative, at least in the case of Shapiro's *ante rem* theory. It posits a rich universe of *ante rem* structures, over and above *in re* set-theoretic (or category-theoretic) structures that may instantiate them. Furthermore, it provides a natural criterion of *mathematical* structures, viz. specified by coherent axioms in the language of a.r. structure theory.

Concerning point (4), *ante rem* structure theory provides explicit conditions of mathematical existence of a.r. structures, much as set theory does for set-theoretic structures. What about the question of indefinite extendability of structures? While it is true that every particular a.r. structure has a proper extension, e.g., via the axiom of power-structures, there are implicitly maximal totalities that arise in a.r. structure theory. For example, an instance of the second-order comprehension scheme implies the existence of a class of *all places* in all a.r. structures, much as von Neumann-Gödel-Bernays (NBG) and Morse-Kelley (MK) each imply the existence of a proper class of all sets, or a class of all ordinals, etc. There is thus an implicit commitment to a fixed universe of a.r. structures themselves, even if they do not constitute a single "super-structure." Again the situation is analogous to that of STS, which recognizes a fixed, absolutely maximal background of all sets. Such are the Platonist foundations of mathematics.

Proponents of SGS have not said much concerning our point (5), reference and epistemic access to *ante rem* structures. The issues are assimilated to the more general phenomena of pattern recognition and abstraction of types based on multiple instances. Shapiro makes an analogy with letters of an alphabet: initially we learn how to read such and produce multiple concrete instances;

sooner or later, we begin speaking and thinking about letters in the abstract, thought of and treated as types rather than tokens. How this is accomplished in detail is left to cognitive psychology and neuroscience. In the case of mathematical structures, presumably our initial steps are taken with respect to small, finite samples of concrete things. We note the phenomena of one-to-one correspondences between different groups of things of the same cardinality (e.g., knives and forks of place settings at dinner tables, etc.), and we learn how to enumerate small collections, establishing relations between cardinalities and order properties. Early on, we realize that we can always "add 1" at any stage of enumerating collections. And at some point, we learn how to speak of and operate with finite collections and numerals while dropping reference to particular features of the objects involved. We are now on the way toward understanding mathematical operations on finite collections, or on the numerals, e.g., taking successors, forming sums by thinking in terms of disjoint unions, iterating summations to introduce products, and so forth. Our first conceptions of finite cardinal and ordinal structures are taking shape. We now realize that we can always extend our counting beyond any point reached. Eventually, we will begin to grasp the notion of the countably infinite; and later, typically with studying some mathematics, we grasp the idea of the minimal closure of an initial small set under an iterable operation. Already, we are coming to understand second-order mathematical content, and eventually even infinite *ante rem* structures will be within reach. Such is Platonist epistemology.

Let us now consider how SGS fares regarding issue (6), respecting the semantics of ordinary mathematical discourse. This is one of SGS's strong suits, as *ante rem* structure theory preserves a face-value reading of mathematical sentences. For example, singular terms are treated as such, with numerals, for instance, referring to numbers as places in the *ante rem* structure of the natural numbers, and similarly with respect to the integers, the rationals, etc. Similarly, predicates are given direct interpretations over a.r. structures accurately modeling ordinary mathematical theories. In short, "What you see is what you get." This straightforward semantics contrasts markedly with the comparatively complex translation schemes associated with STS and, as we shall see, with CTS, MS, and MST (to be examined below).

Turning now to issue (7), resolving paradoxes, SGS follows STS quite closely with respect to the well-known set-theoretic paradoxes, under the rubrics Burali-Forti, Cantor, and Russell. Concerning Burali-Forti, for example, there can be no "largest ordinal" in light of the extendability principle: the height of any *ante rem* structure for ZFC, NBG, or MK set theory is exceeded by many ordinals in proper extensions. Likewise for the Cantor paradox, concerning the putative "set of all sets." According to SGS, "sets" are relative to an a.r. structure, and every such has a proper extension, so the set $s$ of all sets of a structure $S$ is transcended by many sets $s'$ of proper extensions $S'$ of $S$, and so on.

When it comes to the Russell paradox, however, matters are more complicated. Since SGS takes over the standard set-theoretic resolutions, no Russell *sets* are recognized as places in a.r. structures as models of standard set theories. Nevertheless, one might encounter the following construction. Recall that a.r. structure theory recognizes both structures and systems, and that places of systems are like "offices," fillable by other objects. Now if places in systems can be filled by systems, then a Russell-style paradox can be generated, using second-order Comprehension. Define a system to be "self-instantiated" just if it occurs at a place of itself. Then consider the system of all non–self-instantiated systems. Clearly it is self-instantiated just in case it is not, and this is a contradiction.

In the formal structure-theory of Shapiro (1997), this particular construction is blocked. Systems are second-order items, likened to properties or (possibly proper) classes, and places are in the range of the first-order variables. So a system cannot occur as a place (or at a place) in a system. However, *structures* can so occur, since structures are "objects," items in the range of first-order variables. Say that a system $S$ is *self-fulfilled* if it exemplifies a structure $s$ and $s$ occurs at a place of $S$ (or as an object in $S$). Clearly, some systems are self-fulfilled. Consider, for example, the following system, which is like the natural number system, except that 0 is replaced with the natural number structure:

$$\langle \mathbb{N}, \underbrace{1, 2, 3, \dots}_{\omega-\text{many entries}} \rangle$$

With obvious definitions, this exemplifies the natural number structure (i.e., the natural number structure itself has no predecessor, etc.).

The comprehension scheme (as part of the logic) would then deliver a system of all non–self-fulfilled systems. It follows that this system of all non–self-fulfilled systems cannot itself exemplify a structure. If it did, then a contradiction would result. Recall that in the structure theory, systems are in the range of the second-order variables and structures are in the range of the first-order variables. So, clearly, there are "more" systems than structures. As Shapiro (1997, pp. 95–96) put it:

> Is there a structure of all structures? The answer is that there is not, just as there is no set of all sets. Because a "system" is a collection of places in structures (together with relations), some systems are "too big" to exemplify a structure ... [*Ante rem*] structuralism is no more (and no less) susceptible to paradox than set theory, modal structuralism, or category theory. Some

care is required in regimenting the informal discourse, but it is a familiar sort
of care.

It might be added, however, that this more or less standard resolution seems
unmotivated here. Why should there be restrictions on the ability to abstract
from a (coherent) system to its structure? To paraphrase Michael Dummett
(1991, p. 316), to say that if you persist in talking about the structure of all non–
self-fulfilled systems you will run into a contradiction is "to wield the big stick,
not to offer an explanation."[24]

Let us consider now how SGS fares with respect to the outstanding problems
discussed above affecting STS. Recall that the first of these (i) concerned the
fact that STS does not provide a structuralist treatment of set theory itself,
despite the multiplicity of different set theories, all of which form legitimate
subjects of mathematical study. Clearly SGS remedies this, treating set theories
as automatically concerned with *ante rem* structures so long as those theories
are coherent. Sets are not conceived as abstract particulars but, like numbers, as
places in a.r. structures. These vary with the particular set theory at hand,
including non–well-founded as well as well-founded ones, countable as well
as uncountable, "large" in various senses corresponding to large cardinals of all
sorts, structures obeying Gödel's Axiom of Constructibility, as well as structures with large large cardinals (e.g., Ramsey or measurable, etc.) violating that
axiom, and so on. All of these *ante rem* structures are legitimate objects of
mathematical investigation (assuming that their corresponding sets of axioms
are individually coherent), and there is no need to confront questions about "the
(putative) real universe of sets."

Concerning (ii), violations of the extendability principle stemming from the
commitment to a maximal background "real universe" of sets, SGS appears to
avoid this entirely. There are as many real universes of sets as there are coherent
set theories. And the axioms of a.r. structure theory provide for ever more
encompassing universes of set-like objects, in accord with Zermelo's extendability principle. In consequence, justification of certain key axioms, like
Replacement, is readily available, viz. by appeal to our interest in exploring
large structures in the relevant sense, viz. whose ordinals outstrip bijections
with any elements of the universe in question. Along with this comes natural
motivation for at least the small large cardinals. By coherence, there is an a.r.
structure, call it $S$, for MK, for example; by Zermelo's (1930) quasi-categoricity theorem, the height of this structure $S$ must be a strongly inaccessible

---

[24] It also follows that the structure theory cannot treat the universe of structures itself as an object
of mathematical study. Thanks to an anonymous referee.

cardinal (a regular, strong limit cardinal), and it is guaranteed to occupy a place as a "set" in any proper extension of *S*.

Despite all this, however, due to its second-order logical comprehension principle, *ante rem* structure theory does in fact conflict with the extendability principle in its full generality. For example, the theory proves the existence of a class of *all* ordinals of any *ante rem* ZF-structure for set theory. While this class looks like a maximal ordinal itself, no contradiction is implied, as one cannot infer the existence of any successors (nor of a structure isomorphic to this system). Still, it violates the extendability principle outright, thereby compromising the program.[25] It seems that something much like the Burali-Forti paradox is still with the advocate of SGS. It might be added that, as noted, SGS is committed to a system of all sets in ZF-structures. This, too, compromises the extendability principle as applied to SGS.

An advocate of SGS might hold that the system of all systems, or all ZF-structures, is indefinitely extensible (see Shapiro and Wright 2006). But there seems to be some tension between this and the underlying Platonism. Again, if there is a system of all ZF-structures, then why should there not be a structure of this system?

Turning to point (iii), the "bad questions" arising for STS such as the problem of "missing sets" are bypassed entirely in SGS. Power sets are "full" in virtue of the coherence of entertaining them, as classical set theorists do. There is, likewise, no special puzzle about "the nature of the null set," as this is just a non-membered, initial point in *ante rem* structures of well-founded sets, and the axioms for such are presumed to be coherent.

There are, however, some further problems that have been raised for SGS that do not arise for STS. One of these, let us call it (*iv*), due to Jukka Keränen (2001) and John Burgess (1999) (independently), concerns the identity of places in *ante rem* structures, specifically in structures admitting non-trivial automorphisms. Unlike so-called "rigid" structures, like that of the natural numbers or segments of cumulative hierarchies for set theory, which lack any non-trivial automorphisms, the axioms for non-rigid structures, like the integers, or the complex numbers, or geometric spaces, provide for individuals that cannot be identified merely by their structural roles due to the presence of

---

[25] This is very similar to the problem encountered in connection with Zermelo's proposed framework in his great paper (1930). Presented informally, the framework is naturally formalized as $ZF^2C$, second-order ZFC, including the full second-order comprehension scheme for classes and relations. As a result, despite the fact that Zermelo insists that the "set/proper class" distinction is only relative to a model of $ZF^2C$, still comprehension implies the existence of a class of all ordinals of models, of a class of all strongly inaccessible cardinals, and so forth. While no paradox results, this certainly runs counter to the very spirit of Zermelo's proposal, and indeed to the letter of the extendability principle as stated in the framework.

multiple individuals sharing exactly the same roles. For example, there is no structural distinction to be drawn within the additive group of integers, between the positive integers and the negative integers, between the positive and negative imaginary numbers (multiples of $i$ and $-i$) among the complex numbers, or between points of a (homogeneous) Euclidean space related by a symmetry of the space, such as a translation, a reflection, or a rotation. Yet obviously such places in a.r. structures for these theories cannot be identical with one another. There is, technically speaking, a way of resolving this issue, namely by considering these non-rigid structures as embedded within a set-theoretic hierarchy (which is itself rigid). But this seems to run counter to the whole thrust of SGS, which is to treat coherent axiom systems as implicitly defining structures standing on their own, rather than as interpreted within a privileged set-theoretic hierarchy. The most common reply, among advocates of SGS, is to give up the requirement of a criterion of identity for places in a.r. structures, a kind of identity of indiscernibles (see Shapiro 2006a, 2006b, 2008, 2012, and the references cited therein). The debate over this continues.

Let us turn now to a further problem, (*v*), concerning the very intelligibility of *ante rem* structures. One way to put the question is this: Are purely structural relations intelligible in the context of special putatively structural *objects*? This puzzle affects both rigid and non-rigid structures. One slogan of SGS is that places in *ante rem* structures are determined entirely by their structural roles, that is, by purely structural relations. Consider the natural numbers of the putative unique a.r. structure of the natural numbers. Somehow we must first grasp the purely structural relation of *immediate succession* in order to make sense of reference to the individual numbers, $0, 1, 2, \ldots$ But what sense can we make of *the successor relation* if we do not already presuppose reference to the numbers themselves, in terms of which we specify that particular relation? Yet on the SGS understanding of "the numbers," reference to them can only be grasped in terms of their structural roles, and those are given with reference to the relation of "immediate succession"!

We seem to be caught in a vicious circle: access to purely structural relations depends on prior access to the *relata*, whereas access to the *relata* depends on prior independent access to the purely structural relation(s). Can we appeal here to the axioms (i.e., of Dedekind and Peano) to pick out one or the other? Well, no, because there are multiple equally legitimate interpretations of those axioms, as Russell pointed out in his criticism of Dedekind in his early expression of *ante rem* structuralism. What can it mean to speak of *the ordering* of *the natural numbers* as objects unless we already understand what these objects are *apart from their structural roles*? Surely the notion of "next object" makes sense only relative to an already grasped ordering, function, or arrangement of the objects. It might be noted that Dedekind was sensitive to this in that he always was careful to specify a succession function $\varphi$ when referring to a simply infinite system (set) of objects to be identified with the numbers, as he wrote "as set in order by transformation $\varphi$."

Whereas the previous Keränen-Burgess objection granted the relations and raised the issue of how these could determine the objects of non-rigid structures, the present objection questions such talk of relations in the first place, and thereby the very idea of "Dedekind abstraction," to arrive at a.r. structures. One response to this might be to give up the above slogan that the objects in an *ante rem* structure are determined by the relations of the structure. This is problematic in any case, since SGS envisions structures (like some trivial graphs) that have no relations whatsoever (unless identity and non-identity count as relations). So the objects of *these* structures cannot very well be determined by its relations, since they do not have any (non-trivial) relations. A better slogan might be that the objects and relations are somehow determined relative to each other. Neither is primary, either ontologically or epistemologically. The details of this metaphysics are yet to be filled in.

Similar considerations lead to another Benacerraf-style worry. Start with Dedekind's formulation, and consider "the numbers $1, 2, 3, \ldots$" as "set in order by successor function" $\varphi$. Now consider the numbers (say) $2, 1, 3, \ldots$ as just as surely "set in order" by successor function $\varphi'$ defined in the obvious way by permuting the roles of the first two elements. Which of these is the true *ante rem* structure of *the* natural numbers? It seems clear that one choice is just as good and legitimate as the other. They are clearly not discriminated by appealing to the axioms: both are perfectly good models of those. The conclusion is this: *the notion of ante rem structure appears beset with a vicious circularity: such a structure is supposed to consist of purely structural relations among purely structural objects, but understanding either of these requires already understanding the other* (see Hellman 2005 and Assadian 2017 for a similar objection).

Thus, it seems that SGS is ultimately confronted by the problem of multiplicity leading to the Benacerraf objection. Given one *ante rem* structure satisfying the Dedekind-Peano axioms, there are infinitely many others with equal claim to serving as "the natural numbers," and no evident way to say that any one of them serves as the genuine referents of our number words and the relevant relations among them. Indeed, as we already have mentioned, Benacerraf generalized his original point, that numbers cannot be identified with sets due to multiple equally good choices, to the claim that numbers should not be conceived as definite *objects* at all. At first, it may have seemed that a.r. structures provided a basis for such a choice, but now we have seen that that is an illusion. Hyper-Platonist abstraction, far from resolving the problem, leads straight back to it.

It is hard to adjudicate the various intuitions that are invoked in this debate. The SG structuralist does not provide an apodeictic argument to the conclusion that, given coherent axioms, there is a unique *ante rem* structure that satisfies them. The uniqueness is a postulation of the theory, perhaps justified on grounds of simplicity or ontological economy. The fact that, given the existence of a given structure, there are systems that are isomorphic to it, and so have an "equal right" to be called *the* structure in question, does not refute the postulate. Perhaps one can explore the option of giving up uniqueness, but then one cannot insist that mathematical statements be taken at face value. We would no longer have a unique referent for expressions like "the natural numbers" and "the natural number six."

We conclude this section by mentioning one more problem with SGS. It concerns the Coherence axiom, according to which the coherence of a class of axioms justifies our taking those axioms as defining at least one *ante rem* structure. The new primitive, "coherent," may be thought of as the post-Gödelian descendant of "formal consistency," and the Coherence axiom the descendant of the Hilbertian idea that consistency suffices for mathematical existence. But "coherence" is not a formal notion and seems no clearer than the idea of second-order logical possibility. As we shall see in the section on modal-structuralism below, second-order logical possibility serves as a neo-Hilbertian interpretation of mathematical existence. But, unlike "coherence," it is governed by the formalism of (S-5) quantified modal logic, and second-order logical possibility is distinguished from actual existence. From the modal perspective, as we shall see, it seems questionable that mere coherence should suffice for genuine existence (as opposed to mere possibility). Especially in connection with the principle of indefinite extendability of structures (for set theory and topos theory), modality introduces a new degree of freedom useful in avoiding commitments to all-embracing totalities.

As might be surmised, the present authors do not agree on whether the above objections to SGS are decisive. But we do agree that further perspectives should be explored.

## 6 The Modal-Structural Perspective

A good way into the approach of modal-structuralism (MS) (developed in Hellman 1989, 1996) is via Russell, who wrote early on (in Russell [1919] 1993):

> It might be suggested that, instead of setting up "0," "number," and "successor" as terms of which we know the meaning . . ., we might let them stand for *any* three terms that verify Peano's five axioms. They will then no longer be terms which have a meaning that is definite though undefined: they will be "variables," terms concerning which we make certain hypotheses, namely, those stated in the five axioms, but which are otherwise undetermined. . . . our theorems . . . will concern all sets of terms having certain properties. (p. 10)

No sooner had Russell offered this suggestion than he retracted it for what we today would regard as quite spurious reasons (the *definition* did not provide for *existence* of models [this from a critic of the ontological argument!], and you could not account for ordinary counting on the proposed interpretation),[26] and pursued the Fregean absolutist line (cardinal numbers as definite classes of equinumerous concepts or classes), only to encounter various problems and paradoxes leading to the abandonment of classes. In the end, we find this remarkable suggestion concerning logical and mathematical propositions generally:

> We may thus lay down, as a necessary (though not sufficient) characteristic of logical or mathematical propositions, that they are to be such as can be obtained from a proposition containing no variables . . . by turning every constituent into a variable and asserting that the result is always true or sometimes true. . . . logic (or mathematics) is concerned only with *forms*. (Russell [1919] 1993, p. 199)

When we consider that *relations (propositional functions)* as well as individuals count as "constituents" of propositions, then we realize that this criterion is met by formulating mathematics in higher-order logic without constants. In fact, second-order logic suffices. In the case of number theory, for example, an ordinary statement A naturally goes over to a (universally generalized) conditional of the form,

$$\forall R \ [PA^2 \rightarrow A](S/R),$$

in which "$PA^2$" stands for the conjunction of the (Dedekind-)Peano axioms and "$S/R$" indicates systematic replacement of the successor constant with the relation variable "$R$" throughout. (Here "0" can be dropped as it is definable

---

[26] On the eliminativist strategy suggested, counting would be understood, roughly speaking, as indicating, with numerals or related symbols, a one-to-one correspondence between the enumerated items and an initial segment of any progression (or any that there *might be*, on a modalized version). One can even put it constructively: counting provides a means of building a one-to-one correspondence between the enumerated items and a segment of any given progression. On such an account, *standing in such correspondences* plays the role that *membership* plays on the Frege-Russell account (i.e., the enumerated class [or concept] *belongs to* a Frege-Russell number).

from successor.) If A is logically implied by the axioms, this is a truth of second-order logic; if not, the result of replacing "A" with "¬ A" is, in light of Dedekind's categoricity theorem.[27] Even better, taking the predicate "number" into account as suggested, and generalizing, we obtain,

$$\forall X \; \forall R \; [\text{PA}^2 \rightarrow \text{A}]^X (S/R),$$

where the superscript indicates relativization of all quantifiers to domain $X$. Thus, Russell has come full circle, for this is just the general interpretation "through variables" that he had suggested and dismissed at the outset. (But no matter. Russell eventually got there.) It is already a kind of structuralist interpretation, expressing that truths of arithmetic are what hold in any progression whatever. Formulated as it stands, however, it is inadequate, for suppose there *are no* progressions; then all such conditionals are vacuously true, regardless of the content of A. A better plan is to construe the generalization "always true" modally, i.e., as meaning "in any progression there might be, logically speaking," prefixing the last displayed formula with a necessity operator ("□").

$$\square \; \forall X \; \forall R [\text{PA}^2 \rightarrow \text{A}]^X (S/R). \tag{Hyp}$$

Then to avoid vacuity, we may categorically lay down,

$$\blacklozenge \; \exists X \; \exists R \; [\text{PA}^2]^X (S/R), \tag{Cat}$$

affirming the logical possibility of a progression, which of course is compatible with the actual absence of any. ("Hyp" stands for "hypothetical component [of the modal-structural interpretation]", and "Cat" stands for "categorical component.") The same plan works for the real number system, the complexes, for cumulative hierarchies of sets (characterized by cardinality of *Urelemente* and ordinal height), and, indeed, for any mathematical structure categorically characterized by second-order axioms. (For non-categorical theories, as, e.g., in abstract algebra, the result is incompleteness, but that is as it should be. Of course, more specific types of structures, e.g., transitive permutation groups of a given order, etc., can be treated.) So far we have the basic plan of modal-structural interpretations of mathematical theories.

---

[27] The question may be raised here, in what background theory this categoricity theorem is proved. It need not be anything so strong as impredicative set theory. Indeed, as shown in Feferman and Hellman (1995), even a weak axiomatic fragment of weak second-order logic suffices, in which quantification over finite subsets of the domain is taken as given. More precisely, the theorem is recovered in "EFSC," an elementary theory of finite sets and classes of individuals, conservative over Peano Arithmetic.

The background logic is second-order, with quantified S-5 modal logic, without the Barcan formula.[28] However, care must be taken in formulating second-order logical comprehension. In a modal context, unrestricted comprehension leads to intensions, transworld classes and relations. For example, suppose the predicate "planet" is available; then we would generate a class not merely of existing planets, but of those together with "any that *might have existed.*" That is, we would be recognizing *possibilia.* (Notice that this follows even if the class quantifier is understood as a plural quantifier.) In general, we would be quantifying over relations, not merely relating objects in the actual world or in any hypothetical one being entertained, but *across* worlds.[29] In particular, we would generate a universal class of all *possible* objects, and corresponding universal relations among possibilia, directly violating the extendability principle (modally understood, appropriately, as "Any totality there might be might be extended"). Ordinary mathematical *abstracta* seem tame compared to such extravagances; indulging them would deprive MS of much of its interest as a distinctive program. To avoid such commitments, therefore, an extensional version of comprehension is chosen:

(Logical Comp)          $\Box\ \exists R\ \forall x_1 \ldots \forall x_n (R(x_1 \ldots x_n) \leftrightarrow \Phi),$

where $\Phi$ lacks free '$R$' and is also modal free. Note that the universal quantifiers are not boxed. In effect, ($n$-tuples of) individuals form collections and relations only *within* a world, not across worlds. (Officially, of course, neither worlds nor *possibilia* are recognized.)

This leaves us, however, with a level of abstract classes and relations or Fregean concepts as second-order entities. This can be dispensed with, however, by appealing to a combination of mereology and plural quantification, provided the possibility of an infinity of individuals is assumed. This itself can be expressed using mereology and plural quantifiers, for example, by the following:

---

(Ax ∞)     "There are some individuals, one of which is an atom, and each of
            which combined with an atom not part of it is also one of them."

---

[28] That is, the inference from $\blacklozenge\exists x\varphi$ to $\exists x\blacklozenge\varphi$ is not in general warranted.

[29] Note that it is *quantification* over transworld relations that is to be avoided. There is nothing to prevent us from applying *particular predicates* to entertain relations among given objects and "others that there might have been," as when we say innocuous things such as, "There might have been a horse larger than any existing one," etc. This does commit us to possible horses as objects, nor, of course, to worlds containing such beasts.

where an atom is an individual without proper parts. Given this, one can get the effect of ordered pairing of arbitrary individuals,[30] and this effects a reduction of polyadic second-order quantification to monadic, which itself can be interpreted plurally. *Thus, the relation quantifier in the above second-order comprehension scheme can be replaced with a class quantifier.* Also, $\Phi$ can contain the part-whole relation of mereology.

Completing the core system is a "comprehension" scheme of mereology itself, guaranteeing the whole ("sum," or "fusion") of any individuals satisfying a given (non-null) predicate:

$$(\Sigma \text{ Comp}) \quad \exists x \Psi(x) \to \exists y \, \forall z \, (y \circ z \leftrightarrow \exists u(z \circ u \land \Psi(u))),$$

where "$\circ$", "*overlaps*," is defined via "*part of*", $<$, as

$$x \circ y \leftrightarrow \exists v \, (v < x \land v < y),$$

and where in $\Psi$, monadic second-order variables are allowed, free or bound. Thus formulated, MS is ontologically neutral to this extent: any objects whatever may stand in relevant structural relationships so long as it makes sense to speak of wholes, or of pluralities, of them, which is all that this machinery requires. In particular, a physical or spatio-temporal interpretation of "part-whole" is not required.[31]

This framework turns out to be surprisingly powerful. On the assumption of (the possibility of) just a countable infinity of atoms, not only can the modal existence of progressions (or $\mathbb{N}$-structures) be derived, but, by repeated use of plurals and mereology, full, polyadic classical third-order number theory, equivalently second-order analysis, can be recovered. If one postulates the possibility of a continuum of atoms, this can be pushed up to fourth-order number theory. Even second-order is already rich enough to represent vast amounts of ordinary mathematics, yet there is no use of set-membership or even an abstract ontology of classes and relations. Structures and structural relations are gotten at indirectly; set-theoretic and higher-order logical constructions can be encoded by talk of pluralities (invoking pairing for relations), but without actually reifying structures or relations. We have a "structuralism without structures."

We have now provided answers to the questions under (1) for MS in some detail. Concerning (2), assertory axioms, the initial Axiom of Infinity (Ax $\infty$)

---

[30] See Burgess, Hazen, and Lewis (1991).

[31] It is worth noting, for example, that Goodman (1977), who helped promulgate mereology (which he called the calculus of individuals), applied it to sense qualia, themselves conceived as "abstract" in the sense of "multiply instantiable" and not themselves spatio-temporally bound.

and the displayed comprehension schemata form the core system, to which can be added more specific modal-existence claims for particular kinds of structures. That for $\mathbb{N}$-structures is already derivable from (Ax $\infty$) and instances of comprehension, as is even the modal-existence of continua or $\mathbb{R}$-structures, obtained by following well-known classical constructions, e.g. via Dedekind cuts. For *atomic* $\mathbb{R}$-structures, with reals as atoms, however, further postulation is necessary. Similarly further postulates are needed for structures of higher cardinality, e.g., for Zermelo set theory and beyond. It should also be mentioned that the *unicity* (uniqueness up to isomorphism) of $\mathbb{N}$- and $\mathbb{R}$-structures is also derivable in the core system.

As to (3), this version of structuralism is thoroughly eliminativist, as already described.[32] Concerning what distinguishes *mathematical* from other structures, the same second-order logical criterion mentioned in connection with SGS is available (appealing to translation via plurals).

Concerning (4), MS is uniquely explicit in distinguishing *mathematical existence* from ordinary existence. Zermelo spoke of the former as "ideal existence," an idea similar to the notion of logical possibility. By bringing this to the fore, however, MS permits explicit principles of Extendability of (pluralities of) structures (which Zermelo [1930] articulated for models of set theory), but without generating any universal classes of structures or structural objects (as arise in SGS and in a straightforward second-order logical formalization of Zermelo 1930 as well), due to the natural limitations set by extensional second-order comprehension. It simply makes no sense to speak of a collection or plurality of all structures or items in structures *that there might be*. (This builds on the more mundane fact that, on an ordinary understanding of "collecting," you cannot actually collect anything which only might have existed.)

As to (5), MS is distinguished from other versions in eliminating the need for reference to mathematical structures, since it is only possibilities of structurally interrelated objects that are entertained, and these are given by general descriptions. Ordinary designators, e.g., numerals, occurring in everyday use can be accounted for in various ways, e.g., as indicating relevant places in structures, as convenient devices in counting, measuring, and computing, as introduced in mathematical reasoning modulo the assumption of (the possibility of) a structure of a given type, e.g., $\mathbb{N}$-structures, following the logical move of existential instantiation, and so forth. But the usual puzzles (and "bad questions") concerning Platonist reference to *abstracta* do not arise.

---

[32] Of course, the elimination is for pure mathematics, without prejudice to any actual instantiations of structures there may happen to be in the material world.

The real challenge for MS lies in the last question, concerning epistemic access, not in this case to structures themselves, as these are eliminated on this interpretation, but with respect to the possibilities of structures. What sort of evidence can we have for the various modal-existence postulates arising in mathematics, as illustrated above? Of course, we may gain evidence of formal consistency of the associated axiomatic systems (including various strong forms of consistency, e.g., $\omega$-consistency), even if we cannot have a finitistic proof in the central cases of interest. But the second-order machinery of MS is adopted so that *standard* models of theories (e.g., number theory, analysis, set theories) will be describable, and the possibility of these is not guaranteed by formal consistency claims. It seems that we must fall back on indirect evidence pertaining to our successful practice internally and in applications, and, perhaps, the intuitive pictures and ideas we have of various structures as supporting the coherence of our concepts of them. Perhaps that is the best that any version of structuralism can hope for. We will return to this below, when we come to address some specific challenges that have been raised for MS.

Concerning (6), preserving a face-value interpretation of ordinary mathematical discourse, MS appears to depart quite radically from this, at least *prima facie,* with its translation scheme (what we called the hypothetical component of the MS interpretation of a given mathematical theory). This, however, is regarded as a price well worth paying for the gain in ontological economy and, even more important, the resulting dismissal of what it regards as "bad questions," e.g., how do we manage to refer to abstract numbers, sets, and functions, outside the causal order; how do we know that there are not sets "missing" from cumulative hierarchies of sets; and so forth.

It should be stressed, however, that when we take into account the whole of the reconstructed reasoning in the MS framework, there is a case to be made that ordinary usage is in fact respected. (Here MS is similar to CTS, as described in the section devoted to it.) Indeed, Pettigrew (2008) has argued persuasively that the procedure qualifies as respecting an ordinary face-value reading: One begins with a modal-existence postulate on structures of interest; next one applies modal-existential instantiation, and introduces provisional constants, e.g., in the case of arithmetic, "Let an initial object, call it '0,' be given; then by an axiom it has an immediate successor, etc." According to Pettigrew's analysis, these "dummy names" function as what he calls "dedicated variables," behaving logically as free variables, while serving as "temporary names" under a hypothesis (in this case of the possibility of an arithmetical structure). This fixes ideas and one proceeds to prove theorems on these "provisional numbers." Finally, one observes that nothing depended on the identity of such "provisional numbers," so that universal generalization and necessitation are legitimately applied to recover the MS translate of the theorem proved. In this manner, ordinary practice is conceived as occurring, "sandwiched between" the outer logical moves of instantiation and generalization.

When it comes to desideratum 7, resolving the set-theoretic paradoxes, the modal machinery of MS starts earning its keep, as we shall now see. Recall that STS confronted the problem, in connection with the Burali-Forti paradox, of failing to provide ordinal representatives of well-order relations that fail to be modelable as sets, since they have relata of arbitrarily large ranks in the cumulative hierarchy of sets. Indeed, "most" well-orderings turn out to be non-representable in this sense, viz. lack of existence of, say, a von Neumann ordinal $\alpha$ (a set) whose elements are in one-to-one order-preserving correspondence with the items of the given well-ordering. This is unfortunate since the very rationale for introducing ordinal numbers is to provide for canonical representatives of well-order relations (as well as to enable a good ordinal arithmetic). How does MS improve on this?

Recall the Zermelo-Putnam extendability principle, pertaining to standard models of $ZF^2C$, viz. that any possible such model has a proper (end-)extension. As applied to the Burali-Forti paradox, this means that, while at any given model, there are many well-order relations lacking ordinal representatives *in that model*, in any proper extension thereof all those well-orderings gain ordinal representatives. While it is true that, relative to any such proper extension, further unrepresented well-orderings arise, they in turn gain ordinal representatives in still further, more comprehensive extensions, and so *ad infinitum*. Just as the "set/proper class" distinction is relative to a model, so is the distinction between ordinally represented and unrepresented well-orderings. This key relativity goes hand in hand with the *potential* ordinal representability of any well-order relation that there might be. Finally, there can be no ultimate revenge, for it is not even possible to have all possible ordinals or well-orderings co-existing in a single model or world, on pain of contradicting the extendability principle. Furthermore, this comports well with the natural "actualist" understanding of the modal operators: just as there can be no actual collection of all ordinals (or all objects of any sort) that do not actually but might possibly exist, so there could not exist all ordinals in a possible situation that might still not have existed in that possible situation.

As the reader can now articulate, similar resolutions are available in the MS framework of the Russell and Cantor paradoxes. In both cases, the relevant *desideratum* is that any collection or plurality correspond to a set. Without modality, this of course cannot be achieved: relative to any given model, collections or pluralities of sets of arbitrarily high ranks cannot be identified with sets of that given model. However, any such collections or pluralities can be realized as sets in proper extensions of that given model. Furthermore, there can be no ultimate revenge, as it is not possible for all possible collections or pluralities to co-exist, on pain of contradicting the unrestricted extendability principle and violating our actualist understanding of the modal operators.

Finally, regarding issue (8), let us conclude with some remarks on how MS deals with Benacerraf's problem posed by multiple, equally good ways of construing numbers as objects. As we have already stressed, the predicate "is a natural number," and the specific numerals as genuine constants have been eliminated in the MS translation, reflecting the structuralist point that statements "about numbers" are treated as universal generalizations over arbitrary structures of the appropriate type that there might be (logically speaking). The result is that the Benacerraf problem simply does not arise in the MS setting: numbers are simply not treated as objects at all, in accordance with the conclusion that Benacerraf drew from his observations about multiple equally good interpretations. Here the MS resolution is quite similar to that of CTS, more so than to that of STS, where one does employ some specific identification of finite ordinals, even though one refrains entirely from saying that these are what the numbers "really are." In contrast, MS and CTS avoid even this much identification, remaining consistently on the plane of schematic generalization, systematically replacing constants with variables. Such is eliminative structuralism.

Turning briefly to the problems (i)–(v) raised above, it should be clear that none of them affects MS. Set theories are interpreted structurally, and questions about "the real world of sets" do not arise. Multiple structural possibilities are allowed for, including full "power sets," less-than-full, well-founded domains, non–well-founded, etc. Extendability principles are explicitly part of the interpretation of set theory, leading to the small large cardinals (inaccessible, hyperinaccessible of all orders, Mahlo, $n$-Mahlo, etc.). And, as explained, extensional comprehension does not permit recognition of maximal totalities such as that of "all possible structures." Finally, the whole thrust of MS is to avoid postulating special abstract objects, so the puzzles concerning "places" and "purely structural relations" do not arise. The MS route to "abstractness" consists, not in attempting to introduce "featureless objects," but in simply not building into the descriptions of hypothetical structures anything beyond what is of mathematical interest. Benacerraf's puzzle is solved by accepting his conclusion: numbers as objects are officially eliminated, although number-

words can be introduced as aids in computation and reasoning. Finally, regarding (vi), MS is as explicit as any version regarding its assumptions.

The main new problem for MS is reliance on primitive modality (call this
(vii)), analogous to SGS's reliance on the primitive "coherent." One would like
a formal criterion for these notions, but that is not to be hoped for, and the
approach confronts the epistemological questions broached above.

Space permits us to consider here briefly only the core Axiom of Infinity (Ax
∞), which, with the MS machinery as already explained, suffices for the vast
bulk of ordinary mathematics.

The possibility of a countable infinity of objects seems so entrenched in, and
indispensable to, our scientific and mathematical thinking that it is difficult to
argue against the skeptic. Intuitionists who insist on human mental constructions cannot be satisfied as, of course, we have only finite resources to work
with and can only work so fast. From the classical MS perspective, the issue of
supertasks is entirely beside the point. As Hale (1996) recognized, it is the
ready conceivability of situations in which infinitely many mind-independent
objects exist that we naturally appeal to if pressed. Of course, the Platonist can
claim, in a variety of ways, that our present situation is such, since, e.g., a
proposition $p$ exists and, for every proposition $q$, the statement *"It is true that
$q$"* expresses a proposition, distinct from $q$ (Bolzano 1950); or that some object
$o$ is given, and that for any object $a$, the singleton of $a$ exists and is distinct from
$a$ (Zermelo), etc. But nominalists are not deprived of the possibility of infinities
just for not going along with propositions, sets, etc. It is sufficient if, for
example, it could be the case that there is a moment of time and that for
every moment of time there is a later one (perhaps forming a convergent series),
or – as Dummett concedes to be perfectly intelligible – that there be stars such
that, for any one of them, another one exists some further distance away, etc.
Now, in examining the move from conceivability to possibility, Hale (op. cit.)
explicitly distinguishes between requiring that the conceived situation be one *in
which* it *could* be verified that there are infinitely many things – a condition he
(rightly) regards as too strongly verificationist – and requiring rather that the
conceived situation be one *of which* it *can* be recognized that, *were it to obtain,*
there would indeed be an infinity of things. The latter can be satisfied if a
sufficiently detailed description is available from which it can (here, in *our*
world) be inferred that an infinity exists wherever the description holds (even
this is not in general quite necessary, according to Hale), whereas the former
might be quite impossible without supertasks. And it is the latter requirement
that is to govern, according to Hale (op. cit., pp. 137–8). So what, then, is wrong
with the appeal to moments of time, stars, etc., as above? Our descriptions
straightforwardly entail the existence of infinitely many things in those

situations, and Hale seems to grant the possibility that those descriptions could hold. We certainly *can* see the entailment of infinity, right here and now. It appears that, several pages later, however, there is a slide in Hale's discussion back to a verificationist requirement, for he writes that "the imagined situation would have to be given by a description of which we *could* tell . . . that, were it to be satisfied, this *would* mandate acceptance of a theory which entails the existence of a completed concrete $\omega$-sequence" (Hale, op. cit., p. 145, my emphasis), and just prior to this he writes that only *ordinary* tasks are relevant in assessing an imagined situation "as being one of which we *could* recognize that, were it to obtain, a concrete $\omega$-sequence would exist" (ibid., my emphasis).[33] Then, not surprisingly, the MS appeal to situations in which there is a later moment of time or another star further away, etc., will be found wanting. It seems that, after all, we would have to be able to *determine* such things *in the imagined situations*, in a strong sense, e.g., have evidence that could not be explained on a strictly finitistic basis. It seems that the red herring of supertasks is again out of the jar.

A further criticism sometimes levelled against MS concerns its use of second-order logic. There are two interrelated parts to this: first, second-order logic, in its intended sense, is not formalizable; and, second, this reflects the fact that a substantial amount of mathematics is thus presupposed, raising the spectre of circularity. In reply, the first point is correct and is a corollary of Gödel's first incompleteness theorem (in the context of Dedekind's categoricity of the Dedekind-Peano axioms). It is the price paid for the gain in expressive power along with the failure of logical compactness. *In seeking a systematic formulation of structuralism, however, one is not attempting to formalize all of mathematics.* The advantages of explicitness and clarity concerning one's assumptions speak for themselves, despite the unattainability of completeness. Indeed, the open-endedness and extendability of mathematics are reasons enough to forego the latter aim. Moreover, concerning the second part of the criticism, there is no need to insist on an absolute distinction between logic and mathematics, for MS *does not seek a reduction of mathematics to logic* in anything like the

---

[33] Lest it be thought that we are resolving some subtle ambiguity of modal usage in a biased way, we note that, in the ensuing discussion, Hale writes:

[The structuralist] must supply a description – necessarily finite – of a possible situation, no empirically adequate theoretical account of which could avoid postulating the existence of a completed concrete $\omega$-sequence. But any description which could – in the present context – be reckoned unproblematic will perforce mention only finitely many observationally ascertainable facts . . . which an empirically adequate theory must explain. (p. 145)

This then rules out closure conditions involving quantification, e.g., "for any moment of time, there is a later one," on the grounds that its satisfaction is not "observationally ascertainable." The slide back to a verificationist requirement is complete.

traditional sense(s) (e.g., to demonstrate the "analyticity" of mathematics). It should be granted that some core mathematical content must be built into one's primitive notions if structuralism is to be articulated at all. Indeed the notions of "arbitrary plurality" of infinitely many objects and "arbitrary part" of an infinitude of atoms are inherently mathematical, as the work they can do in the detailed development of MS makes clear. Nevertheless, it should be emphasized that no *primitive* notion of "relation" or "function" is needed: as noted above, *monadic* plural quantification, combined with mereology, enables a reduction of *polyadic* second-order quantification, i.e., of a full theory of relations. This is a non-trivial gain *vis à vis* versions of structuralism which presuppose "set-membership" or "function" or "relation." The claim would be that, while far-reaching in their mathematical import, the notion "some of these things," in the plural sense, and that of "part of a whole" of pairwise discrete things, are accessible to us in ordinary contexts and not special to mathematics. Moreover, surely plural quantifiers belong to *logic* in a general sense, even if mereology's status remains moot. Thus, MS could be said to establish a *partial logicism*, less ambitious than the full program to be sure, but significant nonetheless.

A further, somewhat esoteric problem has been raised bearing on the Extendability Principle and the Quasi-categoricity of $ZF^2C$. It concerns the possibility of "metaphysically shy or exclusive objects," objects that possibly exist relative to the actual word but which cannot co-occur in the same world. An example (due to Williamson, cited by Linnebo [forthcoming]) would be the possibility of forming two distinct knives using the same handle with either of two different blades. The results of each assembly are possible separately, but they are not compossible. This complicates the situation regarding the formulation of quasi-categoricity: It just is not the case that a model of (applied) set theory with one of these knives has an extension to a model with the other one. On the other hand, it is crucial for MS to avoid commitment to embeddings that literally map the contents of one possible world to those of another, as those are intensional constructions that depend on the metaphysical posits of possible worlds and possibilia, anathema to the whole MS program. What should one say about this?

One resolution is to weaken the statement of quasi-categoricity to require simply that any model M (of applied $ZF^2C$) be compossible with a proper (end-) extension, M′, which is itself isomorphic with another, N, where N contains the other object non-compossible with one (or more) of the objects of the original M. Here, while M′ is compossible both with M and also with N, it may not be the case that all three are compossible. Still quasi-categoricity has its intended force, which after all requires embeddability only "up to isomorphism." It should not be surprising, however, that the avoidance of intensions exacts a price elsewhere in the framework, in this case, in the need for a more complex formulation of quasi-categoricity. Indeed, a major theme of this study is that such trade-offs are inevitable.

Other criticisms of MS have been offered, especially in connection with applications of mathematics, an issue that space has not permitted us to deal with in this section.[34] In general, structuralists are well-positioned to treat applications since these are naturally understood in terms of full or partial instantiation of mathematical structures by material systems, or, in some cases, just via mappings between these. As an eliminativist version, MS does require some artful manoeuvring to express relevant relationships between material systems and hypothetical objects that merely *might* form mathematical structures, e.g., $\mathbb{N}$-structures, $\mathbb{R}$-structures, or the various spaces of analysis, etc. But the methods worked out in Hellman (1989, Ch. 3), together with improvements from Burgess and Rosen (1997), I believe essentially solve this problem.

## 7 Modal Set-Theoretic Structuralism

### Introduction

The perspective on mathematical structure recounted in this section grew out of original work by Charles Parsons and Øystein Linnebo developing an explicitly modal account of set theory per se. (The detailed version we follow here is based primarily on the work of Linnebo 2013.) Now, since set theory itself has all the machinery needed to develop model theory – as already reviewed in the case of STS – any proposal that succeeds in reconstructing a reasonably rich set theory on a new basis brings with it the resources for a version of set-theoretic structuralism. We saw this in the section above on CTS, category-theoretic structuralism, where the theory there labeled as ETCS ("elementary theory of the category of sets") qualified as an account of mathematical structure, indeed inheriting certain problems, as well as advantages, of STS. Now, while this is one way in which the program of modal set theory (MST) can provide an account of mathematical structures of ordinary mathematics – that is, for the number systems, for analysis, functional analysis, geometry, topology, abstract

---

[34] Resnik (1997, pp. 74–75), for example, has argued that MS cannot treat ordinary scientific applications of probability and statistics, because of the need for abstract objects such as "events" (e.g., possible outcomes of experiments) and numbers. While Field-style nominalism may be threatened by his objection, I believe it has no force against MS, which can readily invoke the possibility of rich enough structures to *represent* or *model* applications of probability and statistics. What matters is not the metaphysical category of the objects in such a model, but rather the (applied) mathematical information they carry, which depends on our stipulations and on structural roles. Numbers in an absolute sense are no more required as values of probability functions than they are for ordinary counting or measuring.

algebra, etc. – it is by no means the only way. And below, we will describe a natural alternative, which allows all these structures from mainstream mathematics to stand on their own, rather than being identified directly with certain sets of an over-arching set theory.

Our principal reason, however, for taking up modal set theory here is to develop the very interesting contrast it presents vis-à-vis both STS and MS. Recall that STS avoids modal logic entirely in its theorizing about mathematical structures, whether for ordinary mathematics or for set theory itself. And recall that MS, far from *combining* modal operators with set-theoretic primitives (set-membership, "class"), actually *eliminates* those primitives in favor of logically quantifiable variables. In so doing, it respects what has been widely considered a strong point of STS, viz. that it succeeds in *replacing* the older notions of necessity and possibility with the more streamlined appeal to set-theoretic *abstracta*. Indeed, the view is widespread that it is actually a kind of "category mistake" to combine modality with purely set-theoretic vocabulary, much as it is a category mistake to combine the language of tense with purely set-theoretic vocabulary. Thus, just as it is illegitimate to say, "Yesterday, the null set was a subset of its own singleton," or "Tomorrow it will be," etc., so it has been regarded as illegitimate to say, "Necessarily (or possibly), the null set is a subset of its singleton," etc. While not necessarily endorsing this view, MS remains neutral with regard to it, in that it eliminates the characteristically set-theoretic vocabulary from its official interpretation of set-theoretic mathematics. This is an essential aspect of MS's structuralist view of set theory. It represents a direct extension of its eliminative program in the cases of the number systems, analysis, and geometry to the realm of set theory. (Indeed, MS's translation pattern, which we called the "hypothetical component" of its interpretations of the number systems, carries over intact in providing a modal-structural interpretation of *bounded* set theory, where the set quantifiers are restricted to sets of a given rank. [Cf. Hellman 1989, Ch. 2 for details.] NB: Bounded set theory suffices for all of ordinary mathematics, so this situation is not surprising.)

Returning to our motivation for considering a perspective on mathematical structure based on modal set theory, this derives largely from reflecting on the indefinite extensibility of models of set theory. Thus, in counterpoint to the view just described, that it is a category mistake to combine talk of sets with modal operators, the proponent of modal set-theoretic structuralism (MSTS) points to the *potential* character of models of set theory as crying out for the re-introduction of modality into modern mathematics. As we saw in the section on MS, the set-theoretic paradoxes are resolved in a compelling way by appealing to the potential nature of "the cumulative hierarchy of sets." So conceived, it is problematic to think of that hierarchy as ever *complete* (or *completed*), for it is always possible to consider ever more comprehensive totalities, without any possible end to such a process. Indeed, the possibility of richer and richer extensions of any given cumulative hierarchy appears evident by directly extending set-theoretic operations – e.g., taking singletons, or unions, or power-sets (or power-classes) – to the hierarchy itself. Thus, *potentiality* is regarded as built into the very fabric of set theory, despite the well-known successes of such theory, both pure and applied, achieved without recognizing it.

Thus, far from deploying modal logic and the logic of plurals to eliminate set-theoretic vocabulary, MSTS combines those logical resources with set-theoretic vocabulary in order to analyze an iterative conception of sets and set-theoretic structures from the ground up.

Let us, then, turn to presenting the leading ideas and principles of modal set theory.

## Outline of Modal Set Theory

Here we follow closely the presentation given in Linnebo's (2013) paper, "The Potential Hierarchy of Sets."

The first step is to settle on a modal logic suitable for the program. It is assumed at the outset that the program is to develop a framework that realizes an iterative conception of sets, which assumes that a cumulative hierarchy results from (our imagined) "set formation" in an ordinally indexed series of "stages," starting with non-sets (urelements, or possibly just the empty set), then taking power sets at successor stages, and taking unions at limit stages. Sets are thus "formed" only "after" their elements have been "formed," i.e., sets are essentially dependent on their members. Thus, a suitable modal logic should capture an accessibility relation that is a partial ordering, viz. reflexive, anti-symmetric, and transitive; and it should be well-founded, i.e., with no infinite descending chains of stages (a consequence of using ordinals to index the stages). Each possible world is such that its domain can be extended to include more sets, and indeed, accessibility is understood such that the domains of accessible worlds can grow, but not shrink. Once a set is formed, it remains available for mathematical operations forever:

$$w \leq w' \rightarrow D(w) \subseteq D(w')$$

where $\leq$ is the accessibility relation.

The next requirement on the accessibility relation $\leq$ is motivated by considering that, if we have two ways of extending the domain of a given world, it should not matter in which order the extensions are made; although it is not required that all possible ways of extending be exercised at once, it is required that combinations of different extensions of a given world be accessible, i.e., that the accessibility relation be *directed:*

$$\forall w_1 \, \forall w_2 \, \exists w_3 (w_1 \leq w_3 \, \wedge \, w_2 \leq w_3).$$

From these basic properties of the accessibility relation, we can identify an appropriate modal logic. The partial-order properties of reflexivity, anti-symmetry, and transitivity imply that the modal logic be as strong as S-4, whose characteristic axiom reflects transitivity, viz.

$$\Box p \rightarrow \Box \Box p,$$

And the property of directedness corresponds to the requirement that

$$\Diamond \Box p \rightarrow \Box \Diamond p,$$

The modal logic that results by adding this latter principle to S-4 is known as S-4.2. To this is added the principle of necessity of non-identity, viz. that

$$x \neq y \rightarrow \Box (x \neq y).$$

Finally, it is known that these properties ensure that the box and the universal quantifier commute in one direction, in accordance with the so-called Converse Barcan formula:

$$\Box \forall x \varphi(x) \rightarrow \forall x \Box \varphi(x).$$

This latter property ensures that, as one moves along the accessibility relation, domains are non-decreasing (never shrinking).

Before moving on, it is worth noting at this point that the modal set-theory being sketched here differs from the MS program, which chooses the modal logic S-5, which is stronger than S-4.2 in respecting the symmetry of the accessibility relation. Thus, in MS it is allowed that domains of accessible worlds may decrease as well as increase. (This is motivated in part by the consideration that proper subdomains of a given model may still be a model of the purely set-theoretic axioms in question, and as subdomains, it seems that they should qualify as accessible.)

Let us turn now to a second, crucial part of the machinery used by MST, the logic of plural variables and quantifiers. This has already been motivated and described above in connection with the modal-structural framework. Here we shall concentrate on the key axioms governing plural constructions and on the ways in which the logic of plurals interacts with the modal logic. First, comes a comprehension scheme for plurals, of the form,

$$\exists xx \, \forall u \, (u \prec xx \, \leftrightarrow \, \varphi(u))$$

where $\varphi(u)$ does not contain $xx$ free. This simply asserts that, for any condition, $\varphi$, there exist exactly the things to which the condition $\varphi$ applies. In most cases, this seems completely tautologous; but it does have the immediate consequence that there exists an empty plurality, that is, that there are exactly the things that, say, are non–self-identical. But since it cannot be inferred from this that any-thing at all actually *is* non–self-identical, it is harmless. Moreover, it is con-venient in that it leads directly to the existence of the null set, once we state the crucial connection between pluralities and sets (coming up momentarily). It is also worth noting that this plurals comprehension principle is highly impredi-cative, in that the condition $\varphi$ may contain bound plural variables. Predicativists of course will find this objectionable, but it is implicitly widely accepted as part of classical mathematical practice. (More will be said below about the relationship between predicative and impredicative mathematics, when we return to consider how MSTS can recover structures of ordinary mathematics, such as those of the number systems and analysis.)

The next step is the framing of some basic principles governing the interac-tion of plurals and the modal operators. First, we observe that being one among some objects, $xx$, should not vary from world to world. Pluralities are thus assumed to be *stable*:

$$u \prec xx \rightarrow \Box \, (u \prec xx) \tag{STB + $\prec$}$$

and

$$u \nprec xx \rightarrow \Box \, (u \nprec xx), \tag{STB + $\prec$}$$

Although the domains of models can grow as we move out along the accessi-bility relation, particular pluralities remain everywhere the same. (This can be generalized to formulas involving plural variables.) But this latter condition actually goes somewhat beyond the stability conditions. Not only does a plurality $xx$ not "change its mind" about any particular object, $u$, as we shift to different worlds, it also never acquires any new objects. That is, pluralities are *inextensible*, as expressed in the following axiom scheme:

$$\forall u(u \prec xx \rightarrow \Box \theta) \rightarrow \Box \forall u(u \prec xx \rightarrow \theta) \tag{InExt-$\prec$}$$

This in effect is a relativization of the Barcan formula (the converse of the converse Barcan formula, explained above) to the condition $u \prec xx$. Just as the Barcan formula requires that domains of accessible worlds not grow, this inextensibility requirement of pluralities requires that the extension of a given plurality not grow as we move to accessible worlds.

Suppose now that we want to require that a particular formula, $\varphi$, be *extensionally definite*, i.e., have the same extension in all worlds accessible from a given world. We cannot express this in the framework of MST by quantifying over possible worlds, as that is forbidden. But we can use modal and plural logic, simply as follows, to say that $\varphi$ is *extensionally definite*:

$$\exists xx \;\square\, \forall u\, (u \prec xx \leftrightarrow \varphi(u)). \tag{ED-$\varphi$}$$

Since the plurality $xx$ remains identical from world to world, so then, by this principle, does the extension of the formula $\varphi$, i.e., the set of all $u$ such that $\varphi(u)$. (In effect, the extensional definiteness of the associated plurality $xx$ is transmitted to the given formula $\varphi$). This will be important below when we consider the nature of sets, for we will want them too to respect "extensional definiteness."

The next step in the development of MST is to establish a translation from non-modal set theory into the modalized one and prove that the translation respects deducibility. Let $\varphi$ be a formula of ordinary, non-modal set theory. Define its modalized (potentialist) translate $\varphi^\circ$ to be the result of replacing all quantifiers in $\varphi$ with their modal counterparts, viz. $\forall$ gets replaced with $\square\forall$, and $\exists$ gets replaced with $\diamond\exists$. Then it is proved that

$$\varphi_1, \ldots, \varphi_n \vdash \psi \;\text{ iff }\; \varphi_1^\circ, \ldots, \varphi_n^\circ \vdash^\circ \psi^\circ,$$

where $\vdash^\circ$ is the deducibility relation of the modal language built from ordinary $\vdash$ along with the axioms of S4.2 and the stability axioms (above) for the modal-plurals theory. Thus, insofar as we are interested in the logical relations between fully modalized formulas set in a background that includes S4.2 and the stability axioms, we may delete all the modal operators and proceed according to the non-modal theory. Also, as Linnebo observes, formulas that are *not* fully modalized are not subject to the above sketched theorem relating $\vdash$ and $\vdash^\circ$, such formulas can be used to exploit the "finer resolution" made available in the modal set theory to shed light on sets that goes beyond ordinary non-modal set theory. (Cf. Linnebo 2013, p. 214.)

We now come to one of the most interesting aspects of MST, viz. the use of the combination of modal and plurals logic to shed light on what can be called "the nature of sets." A key idea is that sets are collections that are, in some sense, *constituted* of their elements. One component of this is the principle of *extensionality:*

$$x = y \leftrightarrow \forall u \ (u \in x \leftrightarrow u \in y) \tag{Ext}$$

This is then officially adopted as an axiom of MST. But there is more to the intuitive idea that sets are constituted of their elements than this, as it is only an "intra-world" principle, whereas we would like to have information about how sets are identified "as we move from world to world," as it were. But this can be explicated in terms of the notion of extensional definiteness, viz. that a set $x$ has the same elements no matter in which world that set occurs. This can be formalized thus:

$$\exists yy \, \Box \, \forall u \ (u \prec yy \leftrightarrow u \in x), \tag{ED-$\in$}$$

i.e., that *sets are extensionally definite*. Consequently, stability axioms hold for sets, as for plurals, and sets, like plurals, satisfy an inextensibility scheme,

$$\forall x \ (x \in y \rightarrow \Box \, \theta) \rightarrow \Box \, \forall x \ (x \in y \rightarrow \theta), \tag{InExt-$\in$}$$

expressing that as one "moves to accessible worlds," a given set does not grow (although new sets may arise). Indeed, the extensional definiteness of sets enables a proof of a potentialist counterpart to the principle of non-modal set theory that bounded first-order formulas are absolute in transitive structures. (For details, see Linnebo 2013.)

The next step is to formulate a principle expressing that elements are "prior to" the set of them. It turns out that a good way to do this is via the *Axiom of Foundation*:

$$\forall x \ (\exists y \ (y \in x) \rightarrow \exists y \ (y \in x \wedge \forall z \ (z \in x \rightarrow z \notin y))), \tag{F}$$

Indeed, this follows from the principle of well-foundedness of the accessibility relation. (A proof could be given in an extension of ZFC that admitted proper classes.)

Thus far, we have seen conditional claims about what sets exist. Can MST formulate and motivate categorical claims of set existence? Linnebo provides a positive answer, in keeping with an iterative conception of sets. This says, intuitively, that given any objects at a stage, there is a later stage at which a (the) set of those objects also exists. This intuitive principle admits a ready formalization in modal logic with the logic of plurals, thus:

$$\Box \, \forall xx \, \Diamond \exists y \, \Box \, \forall u \ (u \in y \leftrightarrow u \prec xx), \tag{C}$$

Indeed, this is a good explication of the Cantorian principle that any "consistent" multiplicity forms a set, where here "consistent" means "co-exist" or "exist together." In effect, this principle (which Linnebo designates "(C)") assumes that, whenever we speak of "some objects" (using plural variables), we are referring to a "consistent multiplicity" in Cantor's intended sense.

As a kind of corollary, the principle (C) can be used to provide a Cantorian answer to the question: what conditions framed in mathematical language define sets? This is the question of which instances of the naive set comprehension scheme are correct:

$$\exists x \; \forall u(u \in x \leftrightarrow \phi(u)), \qquad\qquad \text{(N-Comp)}$$

In the context of MST, this question can be framed in terms of the modalized naive comprehension scheme:

$$\Diamond \exists x \; \Box \forall u \; (u \in x \leftrightarrow \phi(u)), \qquad\qquad \text{(N-Comp}^\Diamond\text{)}$$

The analysis of the nature of sets provides half the answer. According to that, sets are by their nature extensionally definite – the objects that are the members of a set are necessarily its members:

$$\exists yy \; \Box \forall u \; (u \in x \leftrightarrow u \prec yy),$$

So if (N-Comp$^\Diamond$) holds for $\phi(u)$, then possibly there are objects $yy$ that form the extension of $\phi(u)$:

$$\Diamond \exists yy \; \Box \forall u \; (u \prec yy \leftrightarrow \phi(u)).$$

So if a condition $\phi(u)$ defines a set, in that N-Comp$^\Diamond$ holds of it, then it is possible for the condition $\phi(u)$ to be extensionally definite.

Moreover, conversely, if it is possible for a condition $\phi(u)$ to be extensionally definite, then by (C), that condition indeed defines a set (correlated with the plurality comprising its extension). Thus, putting both directions together, we have the satisfying analysis: a condition $\phi(u)$ defines a set just in case it is possible that it be extensionally definite. Some conditions meet this requirement, but some do not, e.g., those that are indefinitely extensible (like " '$u$ is an ordinal," or " $u$ is a set").

Finally, regarding this analysis, the modal machinery clearly plays a key role. Without it, we have no comparably good way of defining extensional definiteness of pluralities, formulas, and sets, and the Cantorian idea of "consistent multiplicity" versus "inconsistent multiplicity" remains inchoate.

With the machinery of MST so far, Linnebo proves that potentialist translates of all the axioms of ZF, except the Axioms of Infinity and Replacement, are provable in MST, taken as the combination of axioms (C), (F), and the axioms asserting the extensional definiteness of $\in$ and that of $\subseteq$, the subset relation.

(The latter assumption is needed, not surprisingly, in order to recover the axiom of powersets.)

In order to derive the potentialist translation of the Axiom of Replacement, a further technical assumption is used, which formalizes the very intuitive principle that if some formula $\varphi$ is extensionally definite, where the extension of $\varphi$ consists of the objects $xx$, and if $\psi$ has as its extension objects $yy$, which are in one-to-one correspondence with the $xx$, then $\psi$ too is extensionally definite. (This is called "extensional-definiteness for Replacement.") In a nutshell, extensional definiteness is a matter of size. As is well known, the Axiom of Replacement gives expression to the informal idea that set-hood itself is a matter of limited size (the doctrine of "limitation of size"). Thus, any plurality of objects that extend arbitrarily high up in the cumulative hierarchy cannot constitute a set. Ordinary ZF proves that every set has a rank, an ordinal level of the hierarchy where the set first appears; and MST thus gives its own characteristic expression of this.

Now, what about the Axiom of Infinity? (Note, incidentally, that the Axiom of Replacement holds in the structure of the hereditarily finite sets. From the perspective of this structure, the first infinite ordinal, $\omega_0$, is a strongly inaccessible cardinal, and cannot be proved to exist.) For this, MST adds one new modal existence axiom, a "modal reflection" principle:

$$\varphi^{\Diamond} \rightarrow \Diamond\varphi \qquad\qquad (\Diamond - \text{Refl})$$

This says that if a condition holds in the potential hierarchy of sets, then it holds already at some world or model. Clearly this implies the Axiom of Infinity – informally, ever larger finite sets go on without end in the potential hierarchy, so, by $\Diamond$-Refl, there must be a world or model containing no largest such finite set. (Linnebo 2013 gives a rigorous proof of this.)

Finally, it is shown that one can translate in the opposite direction, from MST back into non-modal ZF, preserving deducibility, where here MST includes, in addition to the extensional definiteness axioms, the axioms (F), (C), (ED-Repl), and ($\Diamond$-Refl). As an important corollary, we have that MST is consistent, provided that ZF is consistent (cf. Linnebo 2013, Theorem 8.7).

This concludes our exposition of MST, in the most streamlined and well-motivated form known to us.

## Modal Set Theory as a Perspective on Mathematical Structure

Originally, modal set theory was not conceived as a version of structuralism generally, but as a way of formulating set theory to account for the potential character of "the totality of sets" and related, indefinitely extensible notions. Thus, the leading theorists who have developed modal set theory left open certain key questions that any version of structuralism would have to address, involving how to treat structures for the number systems, for the various spaces of mainstream mathematics (metric, geometric, topological, etc.), and so forth. But given the apparatus of MST, several ways of addressing such questions suggest themselves. The most straightforward would be to treat all mathematical structures as sets, with functions and relations defined on them, following the example of STS. However, there are other approaches, more closely aligned with that of MS, especially considering the pre-set-theoretic apparatus of MST, viz. modal logic and the logic of plurals. Let us postpone considering MS-like options and take up first the STS-like option, and how it addresses or can address the eight groups of issues and questions we have been putting to each of the main perspectives considered above.

Concerning (1), the primitives and the background logic, they have been made fully explicit in the previous subsection of this section. Note, however, that, when it comes to the status of functions and relations as objects, the background logic – the logic of plurals and modal logic – by itself is insufficient in expressive power to go beyond monadic second-order logic. Given a means of coding ordered pairs of objects, plural quantifiers allow us to refer to such pairs collectively, and this can provide a theory of relations adequate for reconstructing ordinary mathematics. But the logic of plurals does not provide a pairing function, and adding modal logic does not help. But of course, once the primitives "set" and membership are assumed, we have the full resources of STS at our disposal, and, as is well-known, this includes a very powerful theory of functions and relations, more than adequate for recovering mainstream mathematics.

Concerning (2), the question of axioms, assertoric axioms are clearly indicated in the formal machinery of MST, presented in outline above. Furthermore, concerning (3), structures as sets-cum-functions-and-relations are of course not eliminated. As to what counts as a *mathematical* structure, MST can follow the lead of STS, liberally admitting as "mathematical" any structure describable in the language of pure sets and pluralities. When it comes to urelements, it is a conventional matter whether to count structures involving them as mathematical or extra-mathematical. (Thus, set-theory applied to mechanics or economics, etc., can be taken as dealing with extra-mathematical entities, although the treatment may be highly abstract, or combinatorially complex, calling for powerful mathematics, and, more interestingly, "results" may be arrived at entirely deductively, possibly assuming extra-mathematical axioms (scientific laws, for example). Applying the deduction theorem to these yields purely mathematical results, despite extra-mathematical labels that may enter, as when, for example, a function is labeled a force field or a potential, etc.

When it comes to issues under (4), we arrive at something that differentiates MST from STS, viz. the treatment of indefinite extensibility. As we saw in Section 4, set theory as standardly understood recognizes proper classes in an absolute sense, the class of all sets, the class of all von Neumann ordinals, etc. Precisely here, however, MST parts company from mainstream set theory, as it recognizes the set/class dichotomy only as *relative* to a given model or background set, just as MS does. Thus, like MS, MST embraces its own version of indefinite extendability of set domains, but with this difference: according to MS, extensions of models of set theory, including the axioms of Power Sets and Replacement, are also models, with height extending to the next higher strongly inaccessible cardinal, whereas MST typically countenances small jumps that fall well short of models of those powerful axioms. But no matter: both insist on relativizing the set/proper class distinction, radically breaking with STS. Furthermore, MST can avail itself of the same rationale for avoiding any maximal "world" of sets – the rationale that appeals to the extensional definiteness of pluralities and sets. (You can speak only of collections of co-existing objects, not of merely possible ones relative to the world at which the evaluation is made.) Indeed, the very formalization of extensional definiteness of pluralities and sets is a friendly amendment to modal-structuralism, although MS eliminates the predicate "set" in favor of "object behaving in accordance with the ZFC axioms."

How about the issues under (5), reference and epistemic access to structures? This is an area ripe for further development, as the developers of MST do not speak directly to these topics. There are hints, however, as principles like (C) above relating sets to plurals suggest that access to sets depends on prior access to elements treated plurally. But further work is needed to build on such hints to frame an insightful epistemic theory.

Concerning item (6), respecting a face-value reading of mathematical discourse, MSTS departs from that, much as do its cousins, STS and MS, but with an additional departure: the mathematics of set theory itself is translated with modal operators and the logic of plurals. Thus, MSTS stakes out a position rejecting the view that combining talk of sets with modality, like tense, is a "category mistake." In this respect, MST differs from MS, which is designed to respect this view. As compensation, though, MST regards this is as a price well worth paying in order to respect the indefinite extensibility of sets and ordinals, while at the same time incorporating the membership relation of ordinary set theory. Finally, on this item, MSTS can tell the same story as MS about respecting mathematical practice via the use of "dedicated free variables." As in the setting of MS, this even qualifies as a "face-value reading," depending on the views of the mathematicians engaged in ordinary practice. (Again, see Pettigrew 2008.)

Regarding (7), the matter of resolving set-theoretic paradoxes, the resolutions provided by MS, described above, can be taken over intact. For instance, regarding the Burali-Forti paradox, the desirable feature of having an ordinal representing any well-order relation can be realized by moving to proper extensions of any given domain of sets and ordinals, just as in the MS framework. The fact that extensions according to MST may fall short of being new models of the whole set theory does not affect this potential representability of well-orderings.

What about (8), the Benacerraf problem of multiple, equally good candidates for, e.g., *the* natural numbers? Here MST can simply follow the lead of STS – that is, officially, one eliminates the numerals and the predicate "is a natural number" in favor of a representation of a natural-numbers *structure*, deploying, e.g., von Neumann finite ordinals, or Zermelo's based on the singleton operation, etc. There simply is no claim as to "what the naturals really are."

Let us now turn to the issues (i)–(iii) that we raised in Section 4 as problematic for STS, to see how they affect MST. Recall that (i) concerned the fact that STS does not treat set theory itself from a structural point of view, but rather sees set theory as describing an actually existing realm of sets, despite the fact that there are many mathematically legitimate alternative set theories differing on such basic matters as well-foundedness, the axiom of choice, the ordinal height of the whole hierarchy, and so forth. Does MST confront similar objections stemming from its ontic commitments to a "potential hierarchy" of sets? The matter is moot, for the modal set theorist need not be committed to a unique potential hierarchy; it can construe its axioms as delineating one among several ways of describing hierarchies of sets. Indeed, it is silent on the Axiom of Choice, for example; and it is noncommittal regarding stronger reflection principles, various large cardinal principles, and so forth. Thus, although MST does not offer a structuralist account of set theory in the manner of MS, it is arguably compatible with pluralism regarding "set-theoretic reality."

What about objection (ii) confronting STS, the matter of full power sets versus "missing sets"? Here it seems that MST has the resources to mitigate such problems, for it has the Cantor principle (C), linking potential set existence to pluralities. So long as we recognize a maximal range of interpretation of "any subsets of set S," it seems that we pin down the full power set of S. Of course, one can be a skeptic about plural reference, as well as about "all subsets" of a given set. Ultimately, one probably has to fall back on the practical success of classicist assumptions about plurals and sets. (In philosophy as in life, *c'est la vie*.)

Regarding objection (iii), problems with commitment to a maximal, all-embracing universe of sets, we have already seen that MST, like MS, is designed to avoid any such commitment. It is not at all surprising that there are indeed multiple ways of achieving that.

As indicated above, MST need not follow STS in its treatment of the structures arising in ordinary mathematics. Rather than taking them to be complex sets of various types, MST could adopt the machinery of MS and tell the eliminativist story. In doing so, as we also noted, it needs to add to the modal+plurals logical machinery for a theory of functions and relations, since it would not be invoking set theory for this. Here it can mimic MS and adopt atomic mereology, which yields a theory of relations when it is combined with the logic of plurals. (To remind the reader, the key is the nominalistic reconstruction of ordered pairing, à la Burgess, Hazen, and Lewis 1991.) This works, however, only if an axiom of infinity is adopted in advance, prior to introducing set theory proper, for the theory of relations depends on an infinitude of atoms. But this is no defect, as such an axiom of infinity is certainly needed for the development of ordinary mathematics anyway. The result envisioned, then, is a fairly natural synthesis of MS and MST, with MS governing the structures of ordinary mathematics followed by the development of MST to handle the structures of higher set theory as delineated above. Furthermore, the alignment can be taken a step further if MST adopts – as a friendly amendment – a stronger reflection principle, e.g., second-order reflection due to Bernays, so that proper extensions of domains for set theory are identified as natural models of second-order ZFC, of strongly inaccessible height. Also, as is well known, Bernays reflection implies the existence (here the possibility) of many small large cardinals, all the way up to and including the indescribables. (For a very readable and motivated account of large cardinals, see Drake 1974.)

The main difference that persists between this synthesis and straight MS is over the status of the predicates "set" and "is a member of," with MS eliminating them and the synthesis embracing them. The advantage of MS is that it respects the view that it is a category mistake to describe sets modally, much as it is a category mistake to describe sets temporally. But, in its favor, the synthesis gets by with the most straightforward formulation of the quasi-categoricity of the axioms of ZF(C) (with Choice as a friendly amendment), as it does not face the embarrassment of non-compossibles that MS confronts. (Under the terms of the synthesis, any two pure sets are compossible, as is built into the modal principle of directedness, leading to the modal logic S-4.2.) Thus, as expected, there are trade-offs: no perspective or framework discussed here has all the advantages.

# References

Assadian, B. [2017]. "The Semantic Plights of the Ante-Rem Structuralist." *Philosophical Studies*, https://doi.org/10.1007/s11098-017–1001-7.

Awodey, S. [1996]. "Structure in Mathematics and Logic: A Categorical Perspective." *Philosophia Mathematica*, 4(3): pp. 209–237.

[2004]. "An Answer to Hellman's Question: 'Does Category Theory Provide a Framework for Mathematical Structuralism?'." *Philosophia Mathematica*, 12(1): pp. 54–64.

Bell, J. L. [1986]. "From Absolute to Local Mathematics." *Synthese*, 69(3): pp. 409–426.

Beltrami, E. [1868a]. "Saggio di interpretazione della geometria non euclidea." *Giornale di matematiche*, 6, 284–312. [French trans. in *Annales scientifiques de l'ecole Normale Superieure*, (I)6 (1869), pp. 251–288.1.]

[1868b]. "Teoria fondamentale digli spazii di curvatura costante." *Annuli di mathematica pura ed applicata*, (2)2, 232–255. [French trans. in *Annales scientifiques de l'ecole Normale Superieure*, (1)6 (1869), pp. 347–375.1.]

[1902]. *Opere matematiche*. Hoepli, Milan.

Benacerraf, P. [1965]. "What Numbers Could Not Be," reprinted in P. Benacerraf and H. Putnam (eds.), *Philosophy of Mathematics: Selected Readings (Second Edition)*, Cambridge University Press, 1983, pp. 272–294.

Benacerraf, P. [1965]. "What Numbers Could Not Be," reprinted in P. Benacerraf and H. Putnam (eds.), *Philosophy of Mathematics: Selected Readings (Second Edition)*, Cambridge University Press, 1983, pp. 272–294.

Bernays, P. [1967]. "Hilbert, David," in P. Edwards (ed.), *The Encyclopedia of Philosophy, Volume 3*, Macmillan Publishing Company and The Free Press, New York, pp. 496–504.

Bolzano, B. [1950]. *Paradoxes of the Infinite*. D. A. Steele (trans.), Routledge & Kegan Paul, London.

Boolos, G. [1971]. "The Iterative Conception of Set." In G. Boolos, *Logic, Logic, and Logic*, Harvard University Press, 1998, pp. 13–29.

Burgess, J. P. [1999]. "Review of Shapiro [1997]." *Notre Dame Journal of Formal Logic*, 40(2): pp. 283–291.

Burgess, J. P. and Rosen, G. [1997]. *A Subject with No Object: Strategies for Nominalistic Interpretation of Mathematics*. Oxford University Press.

Burgess, J. P., Hazen, A., and Lewis, D. [1991]. "Appendix on Pairing." In D. Lewis, *Parts of Classes*, Blackwell, Oxford, pp. 121–149.

Cantor, G. [1932]. *Gesammelte Abhandlungen mathematischen und philosophischen Inhalts*, E. Zermelo (ed.), Springer, Berlin.

Coffa, A. [1986]. "From Geometry to Tolerance: Sources of Conventionalism in Nineteenth-Century Geometry." In R. G. Colodny (ed.), *From Quarks to Quasars: Philosophical Problems of Modern Physics*, Pittsburgh University Press, Pittsburgh, pp. 3–70.

[1991]. *The Semantic Tradition from Kant to Carnap*. Cambridge University Press, Cambridge.

Dedekind, R. [1872]. "Stetigkeit und irrationale Zahlen," translated as "Continuity and Irrational Numbers." In W. W. Beman (ed.), *Essays on the Theory of Numbers*, Dover Press, New York, 1963, pp. 1–27.

[1888]. "Was sind und was sollen die Zahlen?," translated as "The Nature and Meaning of Numbers." In W. W. Beman (ed.), *Essays on the Theory of Numbers*, Dover Press, New York, 1963, pp. 31–115.

[1932]. *Gesammelte mathematische Werke 3*, R. Fricke, E. Noether, and O. Ore (eds.), Vieweg, Brunswick.

Demopoulos, W. [1994]. "Frege, Hilbert, and the Conceptual Structure of Model Theory." *History and Philosophy of Logic*," 15(2): pp. 211–225.

Drake, F. R. [1974]. *Set Theory: An Introduction to Large Cardinals*. North Holland.

Dummett, M. [1991]. *Frege: Philosophy of Mathematics*. Harvard University Press, Cambridge, MA.

Feferman, S. [1977]. "Categorical Foundations and Foundations of Category Theory." In R. E. Butts and J. Hintikka (eds.), *Logic, Foundations of Mathematics, and Computability Theory*, D. Reidel, Dordrecht, pp. 149–169.

Feferman, S. and Hellman, G. [1995]. "Predicative Foundations of Arithmetic." *Journal of Philosophical Logic*, 24(1): pp. 1–17.

Frege, G. [1879]. "Begriffsschrift, eine der arithmetischen nachgebildete Formelsprache des reinen Denkens," translated in van Heijenoort [1967], pp. 1–82.

[1884]. *The Foundations of Arithmetic*. J. L. Austin (trans.), 2nd Edition. Harper, New York, 1960.

[1903a]. *Grundgesetze der Arithmetik 2*. Olms, Hildescheim.

[1903b]. "Über die Grundlagen der Geometrie." *Jahresbericht der Mathematiker-Vereinigung*, 12, pp. 319–324, 368–375.

[1906]. "Über die Grundlagen der Geometrie." *Jahresbericht der Mathematiker-Vereinigung*, 15, pp. 293–309, 377–403, 423–430.

[1967]. *Kleine Schriften*. Darmstadt, Wissenschaftlicher Buchgesellschaft (with I. Angelelli).

[1971]. *On the Foundations of Geometry and Formal Theories of Arithmetic*. E.-H. W. Kluge (trans.), Yale University Press, New Haven, Connecticut.

[1976]. *Wissenschaftlicher Briefwechsel*. G. Gabriel, H. Hermes, F. Kambartel, and C. Thiel (eds.), Felix Meiner, Hamburg.

[1980]. *Philosophical and Mathematical Correspondence*. Basil Blackwell, Oxford.

Freudenthal, H. [1962]. "The Main Trends in the Foundations of Geometry in the 19th Century." In E. Nagel, P. Suppes, and A. Tarski (eds.), *Logic, Methodology and Philosophy of Science: Proceedings of the 1960 Congress*, Stanford University Press, Stanford, pp. 613–621.

Goldblatt, R. [2006]. *Topoi: The Categorial Analysis of Logic* (Revised Edition). Dover Publications.

Goldfarb, W. D. [1979]. "Logic in the Twenties: The Nature of the Quantifier." *Journal of Symbolic Logic*, 44(3): pp. 351–368.

Goodman, N. [1977]. *The Structure of Appearance*. 3rd Edition. D. Reidel.

Grassmann, H. [1972]. *Gessammelte mathematische und physicalische Werke 1*. F. Engels (ed.), Johnson Reprint Corporation, New York.

Hale, B. [1996]. "Structuralism's Unpaid Epistemological Debts." *Philosophia Mathematica*, (3)4: pp. 124–147.

Hallett, M. [1990]. "Physicalism, Reductionism and Hilbert." In A. D. Irvine (ed.), *Physicalism in Mathematics*, Kluwer Academic Publishers, Dordrecht, Netherlands, pp. 183–257.

[1994]. "Hilbert's Axiomatic Method and the Laws of Thought." In A. George (ed.), *Mathematics and Mind*, Oxford University Press, Oxford, pp. 158–200.

Hellman, G. [1989]. *Mathematics without Numbers: Towards a Modal-Structural Interpretation*. Oxford University Press, Oxford.

[1996]. "Structuralism without Structures." *Philosophia Mathematica*, (3)4: pp. 100–123.

[2003]. "Does Category Theory Provide a Framework for Mathematical Structuralism?" *Philosophia Mathematica*, 11(2): pp. 129–157.

[2005]. "Structuralism." In S. Shapiro (ed.), *The Oxford Handbook of Philosophy of Mathematics and Logic*, Oxford University Press, Oxford, pp. 536–562.

[2006]. "What Is Categorical Structuralism?" In J. van Benthem, G. Heinzmann, M. Rebuschi, and H. Visser (eds.), *The Age of Alternative Logics: Assessing Philosophy of Logic and Mathematics Today*, Springer Netherlands, Dordrecht, pp. 151–161.

[forthcoming]. "Extending the Iterative Conception: A Height-Potentialist Perspective."

Hellman, G. and Bell, J. L. [2006]. "Pluralism and the Foundations of Mathematics." In S. H. Kellert, H. E. Longino, and C. K. Waters (eds.), *Scientific Pluralism*, Minnesota Studies in the Philosophy of Science, Vol. XIX, University of Minnesota Press, Minneapolis, pp. 64–79.

Hilbert, D. [1899]. Grundlagen der Geometrie. Leipzig, Teubner; *Foundations of Geometry*, E. Townsend (trans.), Open Court, La Salle, Illinois, 1959.

[1900]. "Mathematische Probleme." *Bulletin of the American Mathematical Society* 8 (1902), pp. 437–479.

[1905]. "Über der Grundlagen der Logik und der Arithmetik," Verhandlungen des dritten internationalen Mathematiker-Kongresses in Heidelberg vom 8 bis 13 August 1904, Leipzig, Teubner, pp. 174–185; translated as "On the Foundations of Logic and Arithmetic," in van Heijenoort [1967], pp. 129–138.

[1935]. *Gesammelte Abhandlungen*, Dritter Band. Julius Springer, Berlin.

Keränen, J. [2001]. "The Identity Problem for Realist Structuralism." *Philosophia Mathematica*, 9(3): pp. 308–330.

Kitcher, P. [1986]. "Frege, Dedekind, and the Philosophy of Mathematics." In L. Haaparanta and J. Hintikka (eds.), *Frege Synthesized*, D. Reidel, Dordrecht, Holland, pp. 299–343.

Klein, F. [1921]. *Gesammelte mathematische Abhandlungen 1*, Springer, Berlin.

Lawvere, F. W. [1964]. "An Elementary Theory of the Category of Sets." *Proceedings of the National Academy of Sciences 52*: pp. 1506–1511.

[1966] "The Category of Categories as a Foundation for Mathematics." In S. Eilenberg, et al. (eds.), *Proceedings of the Conference on Categorical Algebra: La Jolla 1965*, Springer, Berlin, pp. 1–20.

Linnebo, Ø. [2013]. "The Potential Hierarchy of Sets." *Review of Symbolic Logic*, 6(2): pp. 205–228.

[forthcoming]. "Putnam on Mathematics as Modal Logic." In R. Cook and G. Hellman (eds.), *Putnam on Mathematics and Logic*, Springer Verlag.

Linnebo, Ø. and Pettigrew, R. [2011]. "Category Theory as an Autonomous Foundation." *Philosophia Mathematica*, 19(3): pp. 227–254.

Mac Lane, S. [1986]. *Mathematics: Form and Function*. Springer, Berlin.

Mayberry, J. P. [2000]. *The Foundations of Mathematics in the Theory of Sets*. Cambridge University Press, Cambridge.

McCarty, D. C. [1995]. "The Mysteries of Richard Dedekind." In J. Hintikka (ed.), *From Dedekind to Gödel: Essays on the Development of the Foundations of Mathematics*, Synthese Library Series 251, Kluwer Academic Publishers, Dordrecht, Netherlands, pp. 53–96.

McLarty, C. [1991]. "Axiomatizing a Category of Categories." *The Journal of Symbolic Logic*, 56(4): pp. 1243–1260.

    [2004]. "Exploring Categorical Structuralism." *Philosophia Mathematica*, 12(1): pp. 37–53.

Nagel, E. [1939]. "The Formation of Modern Conceptions of Formal Logic in the Development of Geometry." *Osiris*, Vol. 7, pp. 142–224.

    [1979]. "Impossible Numbers: A Chapter in the History of Modern Logic." In E. Nagel (ed.), *Teleology Revisited and Other Essays in the Philosophy and History of Science*, Columbia University Press, New York, pp. 166–194.

Parsons, C. [1990]. "The Structuralist View of Mathematical Objects." *Synthese*, 84(3): pp. 303–346.

Pasch, M. [1926]. *Vorlesungen über neuere Geometrie (Zweite Auflage)*. Springer, Berlin.

Pettigrew, R. [2008]. "Platonism and Aristotelianism in Mathematics." *Philosophia Mathematica*, 16(3): pp. 310–332.

Plücker, J. [1846]. System der Geometrie des Raumes in neuer analytischer Behandluungsweise, insbesondere die Theorie der Flächen zweiter Ordnung und Classe enthaltend. W. H. Scheller, Düsseldorf.

Poincaré, H. [1899]. "Des Fondements de la Géométrie." *Revue de Métaphysique et de Morale*, 7, pp. 251–279.

    [1900]. "Sur les Principes de la Géométrie?" *Revue de Métaphysique et de Morale*, 8, pp. 72–86.

    [1908]. *The Foundations of Science: Science and Hypothesis, The Value of Science, Science and Method*. G. Halsted (trans.), The Science Press, New York, 1921, pp. 359–546.

Poncelet, J. V. [1862]. Applications d'analyse dt de geometrie, Mallett-Bachelier, Paris.

Quine, W. V. O. [1986]. *Philosophy of Logic (Second Edition)*. Harvard University Press, Cambridge, MA.

Resnik, M. D. [1980]. *Frege and the Philosophy of Mathematics*. Cornell University Press, Ithaca, NY.

    [1997]. *Mathematics as a Science of Patterns*. Oxford University Press, Oxford.

Russell, B. [1903]. *The Principles of Mathematics*. Allen and Unwin, London.

    (1919)[1993]. *Introduction to Mathematical Philosophy*. Reprint by Dover, New York.

Scanlan, M. J. [1988]. "Beltrami's Model and the Independence of the Parallel Postulate." *History and Philosophy of Logic*, 9(1), pp. 13–34.

Shapiro, S. [1997]. *Philosophy of Mathematics: Structure and Ontology*. Oxford University Press, New York.

[2006a]. "Structure and Identity." In F. MacBride (ed.), *Identity and Modality*, Oxford University Press, Oxford, pp. 109–145.

[2006b]. "The Governance of Identity." In F. MacBride (ed.), *Identity and Modality*, Oxford University Press, Oxford, pp. 164–173.

[2008]. "Identity, Indiscernibility, and *ante rem* Structuralism: The Tale of *i* and –*i*." *Philosophia Mathematica*, 16(3): pp. 285–309.

[2012]. "An '*i*' for an *i*: Singular Terms, Uniqueness, and Reference." *Review of Symbolic Logic*, 5(3): pp. 380–415.

Shapiro, S. and Wright, C. [2006]. "All Things Indefinitely Extensible." In A. Rayo and G. Uzquiano (eds.), *Absolute Generality*, Oxford University Press, Oxford, pp. 255–304.

Stein, H. [1988]. "Logos, Logic, and Logistiké: Some Philosophical Remarks on the Nineteenth-Century Transformation of Mathematics." In W. Aspray and P. Kitcher (eds.), *History and Philosophy of Modern Mathematics*, Minnesota Studies in the Philosophy of Science, Vol. XI, University of Minnesota Press, Minneapolis, pp. 238–259.

Tait, W. [1986]. "Truth and Proof: The Platonism of Mathematics." *Synthese*, 69(3): pp. 341–370.

van Heijenoort, J. [1967a]. *From Frege to Gödel: A Source Book in Mathematical Logic, 1879–1931*. Harvard University Press, Cambridge, MA.

[1967b]. "Logic as Calculus and Logic as Language." *Synthese*, 17(3): pp. 324–330.

von Neumann, J. [1925]. "An Axiomatization of Set Theory." In J. van Heijenoort (ed.), *From Frege to Gödel: A Source Book in Mathematical Logic, 1879–1931*, Harvard University Press, Cambridge, MA, pp. 394–413.

Von Staudt, Karl Georg Christian. [1856–60]. *Beitrage zur Geometric der Lage*. F. Korn, Nürnberg.

Weyl, H. [1949]. *Philosophy of Mathematics and Natural Science*. Princeton University Press, Princeton (Revised and Augmented Edition, Athenaeum Press, New York, 1963).

Whitehead, A. N., and Russell, B. [1910]. *Principia Mathematica 1*. Cambridge University Press, Cambridge.

Wilson, M. [1992]. "Frege: The Royal Road from Geometry." *Noûs*, 26(2): pp. 149–180.

Zermelo, E. [1930]. "Über Grenzzahlen und Mengenbereiche: Neue Untersuchungen über die Grundlagen der Mengenlehre." *Fundamenta Mathematicae*, 16, pp. 29–47; translated as "On Boundary Numbers and Domains of Sets: New Investigations in the Foundations of Set Theory," in W. Ewald (ed.), From Kant to Hilbert: A Source Book in the Foundations of Mathematics, Volume 2, Oxford University Press, Oxford, 1996, pp. 1219–1233.

Cambridge Elements $\equiv$

# The Philosophy of Mathematics

## Penelope Rush
*University of Tasmania*

From the time Penny Rush completed her thesis in the philosophy of mathematics (2005), she has worked continuously on themes around the realism/anti-realism divide and the nature of mathematics. Her edited collection *The Metaphysics of Logic* (Cambridge University Press, 2014), and forthcoming essay "Metaphysical Optimism" (*Philosophy Supplement*), highlight a particular interest in the idea of reality itself and curiosity and respect as important philosophical methodologies.

## Stewart Shapiro
*The Ohio State University*

Stewart Shapiro is the O'Donnell Professor of Philosophy at The Ohio State University, a Distinguished Visiting Professor at the University of Connecticut, and a Professorial Fellow at the University of Oslo. His major works include *Foundations without Foundationalism* (1991), *Philosophy of Mathematics: Structure and Ontology* (1997), *Vagueness in Context* (2006), and *Varieties of Logic* (2014). He has taught courses in logic, philosophy of mathematics, metaphysics, epistemology, philosophy of religion, Jewish philosophy, social and political philosophy, and medical ethics.

## About the series

This Cambridge Elements series provides an extensive overview of the philosophy of mathematics in its many and varied forms. Distinguished authors will provide an up-to-date summary of the results of current research in their fields and give their own take on what they believe are the most significant debates influencing research, drawing original conclusions.

Cambridge Elements $\equiv$

# The Philosophy of Mathematics

Elements in the Series

*Mathematic Structuralism*
Geoffrey Hellman and Stewart Shapiro

A full series listing is available at: www.cambridge.org/EPM

Printed in the United States
By Bookmasters